T0215588

CAMBRIDGE GEOLOGICAL SERIES

EARTH FLEXURES

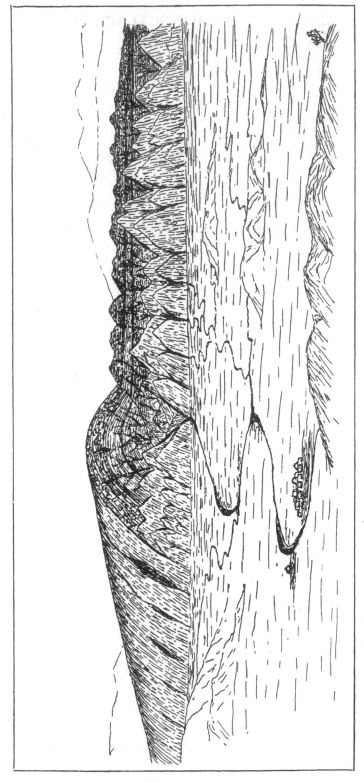

The arch of Kuh-i-Deh Galleh from the plain of Gallol Harkash, Bakhtiari Mountains, south-west Persia.
A monument to law and order in tectonic process.

EARTH FLEXURES

THEIR GEOMETRY

AND

THEIR REPRESENTATION AND ANALYSIS IN GEOLOGICAL SECTION WITH SPECIAL REFERENCE TO THE PROBLEM OF OIL FINDING

by

H. G. BUSK, M.A., F.G.S., F.R.G.S.

ASSOC. MEM. INST. PET. TECH.

CAMBRIDGE
AT THE UNIVERSITY PRESS
1929

CAMBRIDGE
UNIVERSITY PRESS

University Printing House, Cambridge CB2 8BS, United Kingdom

Published in the United States of America by Cambridge University Press, New York

Cambridge University Press is part of the University of Cambridge.

It furthers the University's mission by disseminating knowledge in the pursuit of education, learning and research at the highest international levels of excellence.

www.cambridge.org
Information on this title: www.cambridge.org/9781107663190

First published 1929
First paperback edition 2014

A catalogue record for this publication is available from the British Library

ISBN 978-1-107-66319-0 Paperback

PREFACE

AN all too frequent lack of success in the placing of petroleum test wells in areas which have been the subject of the closest scrutiny has compelled geologists engaged in such work to revise their methods of mapping and section drawing and has forced them step by step to apply geometrical methods where freehand drawing has in the past been deemed to be sufficient. Although this section-drawing work is of general scientific interest over and above its practical application to oil finding, there have as yet been no text books published on the subject, and the matter is rarely referred to in the journals of scientific societies. The most notable paper is one by Sir Thomas Holland in the first issue of the *Journal of the Institute of Petroleum Technologists*, but this is now out of print. It is the first noteworthy attempt to reduce the drawing of earth flexures in geological section to geometrical exactness, and Sir Thomas Holland there represents his curves as a number of straight lines at right angles to the normals through the known dips. Probably a more facile and exact method for the graphic representation of earth flexures is by a number of tangential circular arcs with the normals to the known dips as radii, and it is more along these lines that geological section drawing has of late years developed.

In a text book upon this subject it is difficult to know exactly where to begin, for an aptitude for mathematics and an aptitude for geology are seldom found together in the same individual. It has, however, been thought best to avoid recapitulation of the elements of either science, and the assumption has therefore been made that the reader is fully acquainted with field mapping and section drawing as expounded in such a work as Miss Elles' *The Study of Geological Maps*, published in this series, and that he has some field experience. If he has allowed his early training in elementary geometry to lapse, he may find it necessary to refresh his memory of Smith's *Geometrical Conics*. Care has in fact been taken to exclude anything in the way of repetition of any published relative work.

The analysis of earth flexure forms becomes more intelligible if treated from first principles and as a continuous narrative, and not, as is the all too common practice in text books on other forms of land survey, from the enunciation of a number of rules of thumb. We thus advance by stages from the drawing of the curve of a given bed from the evidence of two proximate dips, to a consideration of that highly complex surface, the "axial plane," which resolves itself in its turn into the study of the locus of intersection of two circles whose radii are augmented or reduced at corresponding rates. As such, to a mathematician, the work may seem elementary, but the application of the simple mathematical problem to geological sections, with the use of geological terms of reference, would puzzle the mathematician, who is not also a geologist.

Chapter II seeks to justify geologically the treatment of earth folds in the manner described.

Much remains to be done. We are still largely in the dark as to how we should interpret the attenuation of the middle limb of a fold from surface evidence, and to the petroleum geologist further research into the matter is of the greatest practical importance; while to the pure scientist, its study with that of the geometry of earth flexures as a whole is as fascinating as any other branch of geology. Geological literature would be greatly enriched by the publication of the many maps and sections, which, on account of competition, now lie in the office drawers of the various great oil companies, and which, unless written up by their respective authors, while they are fresh from their field work, are probably destined to be forgotten.

My thanks are therefore due to the Anglo-Persian Oil Company and to the British Burmah Petroleum Company, in whose service most of these problems have been worked out, for permission to publish that Chapter which deals with illustrations from Persia, Burmah and Egypt.

H. G. B.

1928

CONTENTS

CHAPTER I

INTRODUCTORY

THE study of "Earth Flexures" is one of very wide scope, and it may be approached from a number of points of view. We may consider the "Face of the Earth" as a whole, and endeavour to visualise therefrom the folding of the crust in each successive diastrophic period, as Suess has done: we may approach the subject geophysically, and from the principle of isostatic compensation endeavour to trace cause and effect; or, as has been done in this volume, we may view the subject from the severely practical standpoint of a geologist in the field, whose object is to reproduce, as faithfully as possible from his mapped surface evidence, the form of an earth flexure along any given line of section.

Although there can never be sufficient evidence exposed at the earth's surface for the certain and mathematically accurate determination of what exists below and out of sight, with the growth of the petroleum industry, and the consequent costly field operations of testing and development, there has, during recent years, been a demand for more exactitude and detail in geological methods of survey and section drawing. On this account geology has greatly benefited. There has been greater care and science applied to this branch of the work. Sections, for instance, where the vertical scale is exaggerated, are met with no longer, and there is no one now who would recommend to an oil company an expenditure of many thousands of pounds on a testing and development programme, before being certain that all the evidence that is exposed has been collected, classified and applied strictly according to its value. Accurate contour maps, compiled by means of the most accurate surveys, have replaced the old rough diagrams, produced by map-case and compass, and geological sections no longer rely for their truth (or otherwise) on freehand drawing, but upon careful draughting, so far as the evidence will allow, with geometrical instruments.

This volume is not a treatise on the elements of field mapping and section drawing, ground which has already been covered by many excellent text-books, and a thorough knowledge of the early part of the subject is assumed, nor is it an examination of the many hypotheses that seek to explain the crumpling of the earth's crust, but what it attempts to do is to show how the information supplied from the geological map may be best applied, how geometric methods for section drawing of folds may be used, and to point out those pitfalls that lie in the way of exactitude.

We know really very little about the behaviour of earth folds below surface, and how they enter and die away in the vertical sense; true, we have learnt a little at the expense of costly drilling for oil, but it is not often that sections are exposed to a sufficient depth to tell us whether or not we are right to have made those certain assumptions that are necessary in our sections. Time has proved, however, and there is no doubt whatever about it, that a steady and increasing success is before anyone who will be conscientious enough to apply to all the evidence exposed those methods which are set forth hereafter.

An analysis, for instance, of that very important, but most perplexing part of a fold, the "axial plane" (a misnomer, handed down to us by those who used to draw their sections freehand), will show that it is not a plane at all, but a highly complex curved surface in both the horizontal and vertical sense, a surface, which, provided there is nothing abnormal below ground, may be determined by the mapping of outcrops, in spite of variation in stratigraphical thickness and structural attenuation. Any curve may be expressed mathematically, and there is no doubt that the royal road to rapidity of work in this subject is via the calculus. But there are very few geologists who have the mathematical training for this, and, though the elements may soon be learnt, those expressions, which would have to be applied to the highly complex curves with which we should actually deal, would, probably, be so lengthy and difficult, that it is preferable throughout to work only with graphic methods, and to eliminate anything of such detail that it is not contained within the limits of drawing. There are, probably, also too many sources of error, known and unknown, to make research further than the graphic worth while.

There is to our hand a very simple method of analysing curves or sections of curved surfaces, and it is used by architects, ship designers and by civil engineers in laying out road or railway curves. All curves may be divided into a number of tangential circular arcs, and it is this method of analysis which we shall adopt, and, although we shall be led almost immediately into dealing with "conics," the graphical method can be retained throughout.

By tangential circular arcs we shall find that we can express any flexure from the surface evidence, whether it be a simple anticline or syncline, or, with certain reservations, an overfold with its attenuated middle limb, while much can be done with geometrical instruments, whether thrusting or a known lateral variation in thickness be present or not.

Actual examples from nature will then be analysed, and how the data may be collected in the field by plane table survey will be explained. A consideration of certain specifically folded regions will follow.

It is to be again emphasised that it is no part of this work to theorise on regional folding and how it is brought about, but to deal simply with the evidence as it is exposed in the field, and with what can be induced therefrom regarding that part of any given fold that is hidden.

It is worth while examining in the first instance how far regularity in deposition can be depended upon, and how far irregularities in sedimentary process will affect the problem.

CHAPTER II

GEOLOGICAL PROCESS

I. SEDIMENTATION

WHETHER or not we believe in the permanence of the great ocean basins, or accept tentatively Wegener's striking hypothesis of continental drift, it is generally agreed that sedimentation has, from the earliest times, taken place mainly along the margins of the great continents, which have been subjected to periodic movements of immersion and emersion. Observation has shown beyond this that gradual depression tends to balance sedimentation for long periods, so that, for instance, enormous thicknesses of shallow-water deposits of the same type, many times exceeding the depth of the sea in which they were formed, may occur in great geosynclinal or sagged areas; and that, further, there comes a critical period after such a depression when uplift and folding take place in a high degree, and comparatively suddenly. In many of the present-day great mountain ranges we thus find a very regular succession of sedimentary rocks, showing little variation, both vertically and horizontally, for a considerable thickness and over a considerable area respectively.

It is, however, impossible to generalise on the subject of lateral stratigraphic variation, and, though of course all horizons in a given series must eventually overlap successively against a shore line, the rate of horizontal variation changes with the circumstances under which the deposits were laid down. Variation in a given mapped area may be so small as to be negligible, or of such importance as materially to affect the structural interpretation of the area in section.

For example, in the Tertiaries of Burmah in the Irrawadi basin, where we are dealing with an advancing delta, overlap or change of type may occur in quite a few feet, though in some places thicknesses remain constant over many miles; whereas in Persia, in parts of the Upper Miocene, though there is regional variation, thicknesses and sedimentation generally are invariable over tens of square miles, throughout which sedimentary horizons may be traced with little change of type. In Egypt, in the Upper Cretaceous, regularity in sedimentary process is even better marked.

If we consider sedimentation as it proceeds at the present day, we can readily discern the limits of lateral variation. Sedimentation must, for instance, be fairly constant in rate and in type along the broad strip of the European continental shelf from without twenty miles or so of the coast, and even an estuary of a river such as the Amazon is large enough to ensure regularity over many hundreds of square miles.

There is some tendency, especially in countries covered with forest or cultivation, to explain away any little apparent anomaly in mapping by stating vaguely that there is "lateral variation," where it has afterwards been shown that such an anomaly has resulted through errors in mapping across poorly exposed tracts. "Lateral variation" is in point of fact no explanation of a difficulty, unless it can be shown quite clearly in what direction and in what manner this variation occurs, and it must even then be viewed with suspicion, unless it tallies properly with the general geological history. As an explanation of a difficulty it can sometimes be reduced to an absurdity, where it can be shown that the angle between the converging horizons is unreasonable or impossible in nature.

The variation of a deposit, and whether or not it will have to be taken into account in section drawing, is therefore a question of degree, and depends in part upon the scale of the survey. Thus in Fig. 1, in Region A, an area of, say, fifty square miles, situated

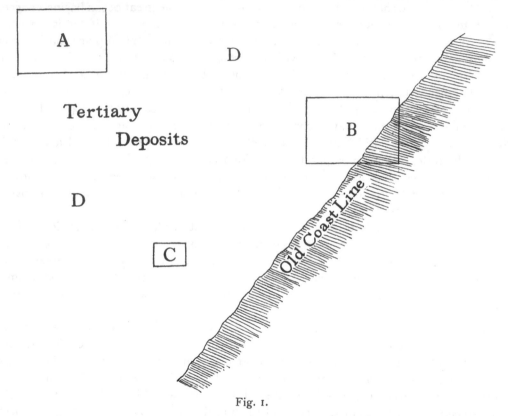

Fig. 1.

fifty miles from the old coast line, variation will hardly be present in such a degree as to affect the map. In Region B, against the coast line, there will be a shore line facies of the beds represented in Region A, and variation in thickness is certain to be present. Region C, to be mapped on a large scale, though situated nearer to the coast line than Region A, is of so small an area, that, though variation will be present, it will hardly be noticeable upon that particular map, while, if we were to survey on a small scale the region covered by the whole figure (Region D), variation both in type and thickness will enter into our sections.

It is to be noted that variation of thickness may be of two kinds; firstly, variation towards a shore line by overlap when two horizons may meet one another, and, secondly, variation by differences in rate of sedimentation, when two horizons may merely be brought nearer to or further from one another.

In general it may be said that in most oilfield regions, even those located in a series which may be said to be very variable, there is usually constancy sufficient for the purpose of mapping horizons over the number of square miles requisite for section drawing, and that, when variation occurs sufficiently to affect the thicknesses in a section line, its exact effect is determinable.

Sedimentary rhythm. Shallow-water clastics often show vertically a "sedimentary rhythm," that is, in a series of alternating clays and sandstones, for instance, the sandstone bands may be found to be separated by nearly equal thicknesses of clay. The reason for this rhythm is not easily found, though it is probably sometimes seasonal; but rhythm is so common that it must be the result of normal process in sedimentation, and not of diastrophic change, uplift and depression. However this may be, in such a series of alternating beds, it is often possible to trace individual time horizons represented by, say, sandstone bands, over very large areas.

Horizon indices. During geological survey, horizons may be dated very simply by taking a single well-marked horizon and indexing the others above and below it by figures representing the numbers of hundreds of feet, at which the other mapped horizons occur respectively, above or below that horizon, the minus sign being used to show that the horizon dated is below the controlling horizon. Thus, if L be the basal bed of the lower group of a certain series, conformable throughout and to the beds immediately below it, then L^6 represents a horizon 600 feet above L, and L^{-4} a horizon 400 feet below L.

When several different localities, say x, y and z, are being mapped at the same time, and it is uncertain whether the L horizon is identical in each of them, it is best to demarcate the horizon mapped in the first locality by Lx, in the second by Ly, and in the third by Lz. Should it be found, when the whole map is complete, that the horizon lines do not meet but interdigitate, then adjustments can be made afterwards by interpolation, and checked by going over the same ground again.

This method of indexing is not only extremely useful for reference, when, for instance, a statement that a certain fossil was found at horizon Lx^{-6} conveys exactly its stratigraphic and geographic position, but in constructing sections it is a useful aid to the memory regarding what horizons are to be drawn in below surface. It is also a reminder that thicknesses must be constantly checked during mapping. Thus in a certain group, whose base is L, in an area which we may call x, there are two well-marked horizons, which have been mapped, Lx^4 and Lx^7; and below L, which is only just exposed in area x, in another area, y, there are two well-marked sandstones Ly^{-4} and Ly^{-8}. In area x, after drawing our section to show Lx^4 and Lx^7, we may extrapolate below surface the two horizons Ly^{-4} and Ly^{-8} at 400 and 800 feet below L respectively. It is of course legitimate to show time horizons at any distance below L, but the expressions Ly^{-4} and Ly^{-8} denote that by drilling we should expect to meet the same sandstone horizons as occur in area y, at 400 and 800 feet below L respectively. Should there be proved variation, say a thickening of 10 per cent., from area y to area x, we may draw in the same time horizons as occur at y at 440 and 880 feet respectively below L. We may now express these horizons as Ly^{-4} ($x^{-4\cdot4}$) and Ly^{-8} ($x^{-8\cdot8}$), denoting that we expect to find the horizons that occur in area y at 400 and 800 feet below L respectively, at 440 and 880 feet below L in area x.

This method of annotation may in point of fact be modified to meet almost any local requirement, and its usefulness as a general check on the work is soon demonstrated.

In a new area, that has never before been mapped, the observer will of course insert as much of the evidence as he can include in the scale adopted, and with this aspect of the subject we shall be dealing in Chapter V. But it cannot be too strongly emphasised here that, for the accurate delineation in section of earth folds, the stratigraphy of any area must be very fully worked out, and variation, if any, must be determined both in its amount and in its direction.

II. FOLDING

Tectonic terms in geology are applied very loosely, and for the purposes of the present work it is necessary to have more exact definitions for some of those that are commonly employed.

We are met at the outset with difficulties arising from certain conventions, which have been in common use ever since geology became a science. The terms "anticline" and "syncline" are always referred to the plane of the horizon, but there is no particular reason why this should be so. Thus an anticline and a syncline may be defined as earth flexures both of whose respective limbs intersect the plane of the horizon, and unless this condition is fulfilled the flexure ceases to be the one or the other; but it can be termed, as we have done here, either an "anticlinal bend" or a "synclinal bend." For oil work, the horizon as a plane of reference has a useful significance, since oil accumulation obeys hydrostatic laws, many of which are dependent upon gravity, which acts in a direction perpendicular to the horizontal plane; but it must always be remembered that this plane of reference is purely a conventional one.

The "crest" and "trough" of an anticline and syncline respectively are referred also to the horizontal plane, but there is some confusion as to what is meant by the "axial plane" and the "apex." The definitions we have adopted are given below, and it will be noticed that the crestal plane and the axial plane, and the crest and apex are rarely coincident. The definition given for the axial plane is not entirely satisfactory, as it can hardly be applied where the crestal area of a fold in section approximates to a semicircle, and it cannot be applied at all where there is attenuation of the middle limb between an anticline and its adjacent syncline.

It seems inevitable that, in the analysis of earth folds geometrically, when we are attempting to apply an exact science to one which has hitherto been treated as inexact, there cannot fail to be occasionally a certain looseness of phrasing. With this premise, the following are definitions of the commoner tectonic terms used in this work, and some of them, be it noted, can only be applied to "competent folds" (p. 9), or folds in which the time horizons remain equidistant from one another. The term "horizon" is, of course, applied indifferently to either a geological time horizon or to the geographical plane of the horizon; which is intended is determinable from the context.

DEFINITIONS

1. The "axial" or "apical" plane of a fold is a surface so disposed within it, that any point upon that surface is equidistant from either limb of the fold, as defined by any particular horizon (Fig. 2).

2. The "apex" of a fold is the line of intersection between the axial plane and either that of the horizon or the ground surface (Fig. 2).

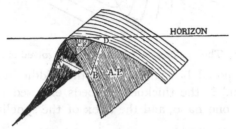

Fig. 2. Defs. 1, 2, 3, and 4. Showing the axial plane (*AP*) and the crestal plane (*CP*). Any point on the axial plane is equidistant from either limb of the fold, i.e. *AB=BD*.

3. The "crestal plane" of an anticline is a surface so disposed within it, that it cuts any horizon in the fold along the line in which that horizon lies horizontally. The crestal plane rarely coincides with the axial plane, but often does so very nearly (Fig. 2).

4. The "crest" of an anticline is a line at ground surface within the anticline, along which all horizons lie horizontally. The crest rarely coincides with the apex (Fig. 2).

5. The "trough" of a syncline is a line within the syncline at ground surface, along which all horizons lie horizontally.

6. An "anticlinal bend" is any relatively sharp flexure in an anticlinal sense, that is, where the beds of the one limb dip gently towards the apex, and the beds of the other dip relatively steeply away from it (Fig. 3). An anticline is merely a special case of an anticlinal bend, that is, where two limbs of an anticlinal bend intersect the same horizontal plane.

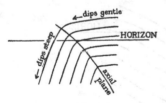

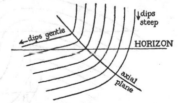

Fig. 3. Def. 6. The anticlinal bend.   Fig. 4. Def. 7. The synclinal bend.

7. A "synclinal bend" is any relatively sharp flexure in a synclinal sense, that is, where the beds of the one limb dip gently away from the apex, and the beds of the other dip relatively steeply towards it (Fig. 4). A syncline is merely a special case of a synclinal bend, that is, where two limbs of a synclinal bend intersect the same horizontal plane.

8. The thickness of exposed beds involved in the middle limb between an anticline and a syncline is the thickness of beds exposed between the crest of the anticline upon the one hand, and the trough of the syncline upon the other (Fig. 5).

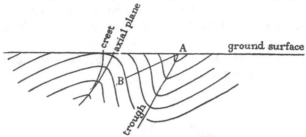

Fig. 5. The thickness of exposed beds involved is *AB*.

9. The thickness of exposed beds involved in the middle limb between an anticlinal bend and a synclinal bend is the thickness of beds exposed between the apex of the anticlinal bend upon the one hand, and the apex of the synclinal bend upon the other (Fig. 6).

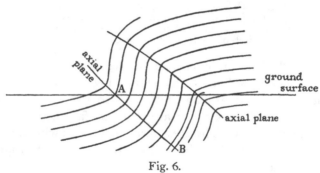

Fig. 6.

10. The amplitude of a given fold is measured by the distance between a horizontal straight line tangential to a certain horizon in the anticlinal part, and a horizontal straight line tangential to the same horizon in the adjacent synclinal part, where that horizon gives a maximum value for the amplitude, as above defined (Fig. 7).

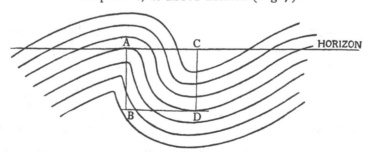

Fig. 7. Def. 10. The amplitude is *AB = CD*.

The degree of magnitude of a fold varies directly as the product of its amplitude and the thickness of the exposed beds involved therein.

COMPETENT AND INCOMPETENT FOLDING

Of the exact process that occurs in a rock during its folding very little is known. American writers divide folds into two classes, those in which horizons maintain parallelism being termed "competent folds," and those in which each horizon maintains a similar curvature being termed "incompetent" or "similar" (Figs. 8 and 9). There may, however, be found in nature every degree of competency and incompetency; an anticline which may be competent in regard to its gentle limb, may by attenuation become incompetent in regard to its steep limb.

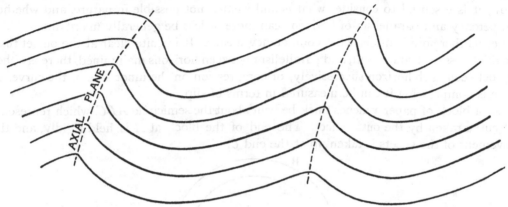

Fig. 8. Competent or Parallel Folds. There is no thickening or thinning of the beds, and the horizons remain parallel. The axial planes of the anticlines are directed away from the steep or middle limb, and the folds tend to die out at depth.

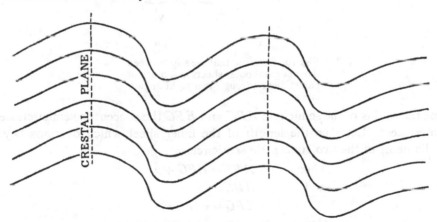

Fig. 9. Incompetent or Similar Folds, where each horizon has a similar curvature. There is a thinning of the middle limb, and the crestal planes of the anticlines are little inclined or remain vertical. There are no readily defined axial planes. The folds do not tend to die out at depth.

It may readily be seen that incompetency implies structural variation in thickness in a fold, a thinning on the limbs and a thickening on the crests and in the troughs, with, if similarity is maintained, no dying out of the fold at depth, while competency and parallelism, on the other hand, will bring about a gradual dying out of the fold at depth. Competency also implies a sliding of the horizons, the one over the other, in the same manner as occurs when a block of paper, held rigidly at one end, is bent.

In most modern standard works on geology, section lines are carried to too small a depth to show whether a fold is competent or incompetent*, but where they are actually so carried, a rough competency is generally shown†, though this does not necessarily imply that the authors had this in mind at the time that the section was drawn‡.

If, for the purposes of oil finding, or for scientific reasons generally, it may be necessary to give as accurate a delineation of a fold beneath surface as the geological evidence will allow, it is essential to consider what is and what is not possible in nature, and whether competency and parallelism of horizons can more or less be generally maintained.

Let us therefore consider what competency means. It is quite clear at the outset that, if a thickness of strata is bent, and parallelism between horizons maintained, there is either slip between each horizon successively, or compression on the inner side of the curve. If there is compression it can be measured in terms of slip.

Let a block of paper x inches thick be bent about the semicircle ABC, which represents the curve taken by the outer sheet. The end of the block at A is held rigidly, and the movement of the sheets is taken up at the end C.

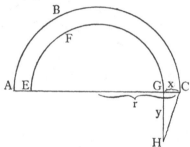

Fig. 10. Showing that the slip between the inner and outer sheets in a block of paper, bent through two right angles, is πx.

Let r be the radius of the semicircle ABC, and EFG the concentric semicircle described by the inner sheet. Let y be the length of the inner sheet, which overlaps beyond the common diameter of the two concentric semicircles.

$$\text{Then} \qquad ABC = EFG + y,$$
$$\text{and} \qquad ABC = \pi r,$$
$$EFG = \pi(r - x).$$
$$\text{Hence} \qquad \pi r = \pi(r - x) + y,$$
$$\text{and} \qquad y = \pi r - \pi(r - x).$$

* Cf. at random, Lake and Rastall, *A Text Book of Geology*, 1910, pp. 332, 446; Chamberlain and Salisbury, *Geology*, 1909, vol. I, p. 503.

† Cf. Lake and Rastall, *op. cit.* pp. 186, 189.

‡ Cf. also Jukes Browne, *The Student's Handbook of Stratigraphical Geology*, 1912, pp. 167, 233. Also many sections in Pascoe's *Oilfields of Burmah*.

Then it is clear that if x is constant, y is constant for any value of r, and

$$y = \pi x.$$

The student can work out actual examples and show that this is so.

Furthermore, bearing in mind that the length of an arc of given radius subtended by any angle between the radii is $\pi r \theta$, where θ is the angle expressed as a fraction of two right angles, it can be shown that y is independent of the fact that the curve taken up by the paper is a semicircle. It may actually be any curve, provided that the sum of the results is a bending through two right angles.

Applying this to a known thickness of beds in a geological section, we may say that:

In any group of strata curved through two right angles, that is, bent back upon itself, the maximum movement between the layers need at no time be greater than π multiplied by the thickness of beds involved.

It may be seen at once how unimportant a consideration this is. Take, for instance, a group of beds 10 feet thick, bent through two right angles in the course of a mile. The total slip of the inner bed relative to the outer will be 10π, or roughly 30 feet. Thirty feet in 5280 feet means ·00056 feet in one foot, or ·007 inches per foot. And such a degree of slip in nature is not only possible, but would be unnoticeable in its effects. Competency means nothing more than this.

Some authors believe that, as there is tension on the outer curve of an anticline, while in the inner curve there is compression, folding movements are taken up by actual compression and attenuation of the beds, but where there are thin limestones or sandstones interbedded with shales, it is quite as likely for the force exerted in bending the strata to be relieved by slip between the shales or along the plane separating the shales from the harder rocks, to the small extent that, as we have shown, will meet the case. When in deeply dissected mountain ranges sections of considerable depth may be observed and measured, and where the folding is of that commonly known as the "Jura" type, that is, not very compressed, parallelism or competency is undoubtedly the most common form adopted, and it is only when folds become compacted in a high degree that parallelism is departed from, and there is attenuation of the steep limbs.

We have dealt with the bending round of a given thickness of beds through 180 degrees in, say, an anticline. If we take the complementary syncline as well, it is easy to see that the slipping movements cancel one another.

For in Fig. 11, the slip of the sheet DEF is πx in the direction FJ, and the slip of the sheet IJC is also πx in the direction CB,

for the curve $ABCJI =$ the curve $DEFHK$ in length.

In nature, of course, the slip is distributed throughout the curve, and is not necessarily concentrated in the middle limb, as shown.

Laboratory experiments in the folding of material under lateral pressure fail in that, although thickness and area may be reproduced to scale, the strength of the materials used and their mass cannot. Thus, if a model is constructed to a scale of 1:250,000, it would seem logical that the strength of the materials with which it is built and their mass should be in somewhere near the same proportion, a condition which it is impossible to apply.

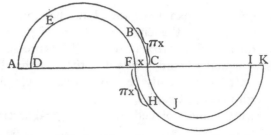

Fig. 11. Showing that the aggregate slip between the inner and outer sheets of a block of paper, bent through two right angles and then bent back again through two right angles, is nil.

A good deal can, however, be learnt from such experiments. Bailey Willis* showed that, as pressure is applied to a model constructed of heterogeneous layers, the more massive layers retain their parallelism or competency for the longest time, while the soft layers rise almost immediately into composite, similar or incompetent folds. Classification in the field is, however, extremely difficult, as competency is really a question of degree. What we are concerned with chiefly is that attenuation of the middle limb or irregularities of that nature are always brought out in field mapping.

The observer in the field must satisfy himself from his map whether competency is maintained or not, and it can be shown from published maps, as well as from numberless maps constructed at the instance of the various oil companies, that competency or parallelism is very common.

We thus have the principle of competency as a starting point from which we may construct sections from the evidence supplied by the map. Complications brought about by incompetency and attenuation render the problem of section drawing less certain, but the general method of work outlined in the following chapters may still be applied with certain modifications.

* Leith, *Structural Geology*, 1914, p. 111.

CHAPTER III

THE GEOMETRICAL CONSTRUCTION OF EARTH FLEXURES IN GEOLOGICAL SECTIONS

W̲E have shown that competency and parallelism in earth folding is, taken as a whole, a general rule, though modification by the attenuation of the middle limb between highly asymmetric folds is not uncommon. Where competency is maintained, the evidence of the geological map is sufficient for the exact delineation of the fold from surface to the depth at which it ultimately dies out, but incompetency introduces factors at depth, which may not be indicated by the surface evidence.

The geometrical constructions for competent folding which follow have in practice given some very remarkable results, and it is quite certain that when the map and section is to be put to the test of the drill, they should always be used. They depend upon the geometrical relationships of tangential arcs, and are based on the following well known proposition:

Prop. 1. **The centres of any two circles that are tangential to one another lie upon a straight line, which passes through the point of contact, and which is at right angles to the common tangent at that point.**

Let the two circles whose centres are at O^1 and O^2 touch at A (Figs. 12 and 13).

Join O^1A and O^2A.

Draw AB, the common tangent to the two circles.

$$\angle O^1AB = \angle O^2AB = \text{a rt. angle};$$

∴ the points $O^1 A O^2$ are in the same straight line.

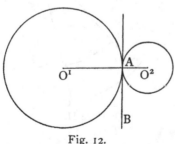

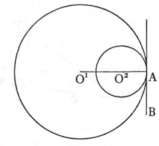

Fig. 12.   Fig. 13.

The constructions which follow depend also on the fact that any curve can be expressed by a series of tangential circular arcs. In geology the complexity of the curve by which a fold can be reproduced in section is limited by the number of dips that can be obtained and mapped, for the tangent to any point on the curve of a horizon in a competent fold is parallel to the dip, that lies at the point of intersection with the ground surface, of the normal to the tangent of the curve (Fig. 14).

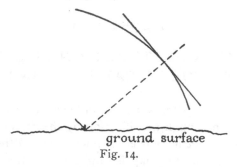

ground surface
Fig. 14.

We may now proceed to construct that part of any competent earth fold which is controlled by two consecutive dip readings.

Prop. 2. **Given any two consecutive surface dip readings in the same direction, it is required to draw the curve taken up by the strata between them, and to measure the thickness of beds exposed.**

Let A be the first dip reading, and B the second.

Draw OA and OB, normal to the dips at A and B respectively, the normals intersecting at O.

With O as centre and OA as radius, draw the arc AC.

With O as centre and AB as radius, draw the arc BD.

Then $AD = CB =$ the thickness of the beds exposed between the dips A and B, and DB and AC are the curve taken up by the strata, in Case I anticlinal, in Case II synclinal.

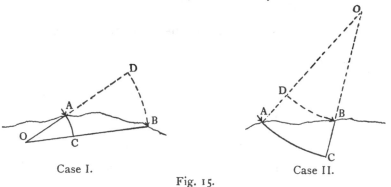

Case I.   Case II.
Fig. 15.

It will be seen from this that the curve taken up by the strata between the two consecutive dips, A and B, is determined by a series of concentric arcs, whose common centre is at O, that is, at the point of intersection between the normals to the two dips in question. It will be noted also that the length of the arc, controlled by the centre O, diminishes as O is approached, until, when this point is reached, it vanishes.

When two consecutively exposed dips occur in opposite directions, it is clear that an anticlinal or a synclinal axis is interposed between them.

Prop. 3. Given any three consecutive dip readings in the same direction, and progressively diminishing or increasing, it is required to draw the curve assumed by the strata, and to measure the thickness of beds exposed between the two outer readings.

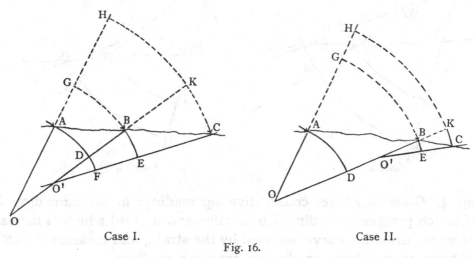

Case I.   Case II.

Fig. 16.

Let A be the first dip reading, B the second and C the third.

The normal to the first dip reading intersects the normal to the second at O, and the normal to the second dip reading intersects the normal to the third at O'.

With O as centre and OA as radius, draw the arc AD,
meeting the radius OB at D.

In *Case I*, with O' as centre and $O'D$ as radius, draw the arc DF,
meeting $O'C$ at F,

and with the same centre and $O'B$ as radius, draw the arc BE,
meeting $O'C$ at E.

Then ADF and BE are the curves assumed by the horizons out-cropping at A and B respectively,

and FC is the thickness of beds exposed between A and C.

It should be noted that the arcs AD and DF are tangential to each other, and that, if that part of the structure, which lies above ground level and has been denuded, is completed,

then GB and BE are tangential,
as also are HK and KC.

For $O'OB$ are in the same straight line.

In *Case II*, the radius $O'C$ intersects OB between D and B, consequently the arc whose centre is at O' is, in that position, cut out. The thickness exposed must therefore be measured along the radius OH or the radius OK, and is $AH = DK$.

In the two cases given the curves are anticlinal.

We give below similar curves, which are in the synclinal sense.

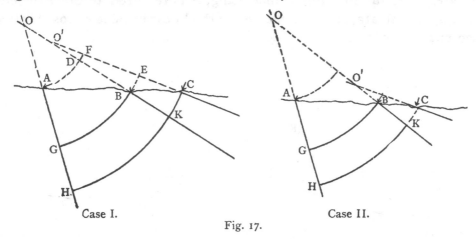

Case I.   Case II.

Fig. 17.

Prop. 4. **Given any three consecutive dip readings in the same direction, two of which progressively diminish in value and a third which is increased or vice versa, draw the curve assumed by the strata, and measure the thickness of beds exposed between the two outer dip readings.**

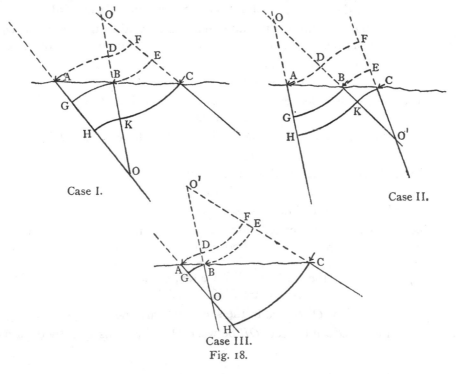

Case I.   Case II.

Case III.

Fig. 18.

The construction is the same as in Prop. 3.

Let A be the first dip reading, B the second and C the third.

The normal to the first dip reading intersects the normal to the second at O, and the normal to the second dip reading intersects the normal to the third at O'.

With O as centre and OA as radius, draw the arc AD, meeting the radius OB at D.

With O' as centre and $O'D$ as radius, draw the arc DF, meeting the radius $O'C$ at F,

and with the same centre and $O'B$ as radius, draw the arc BE, meeting $O'C$ at E.

Also with O as centre and OB as radius, draw the arc BG, meeting the radius OA at G.

Again with O' as centre and $O'B$ as radius, draw the arc BE, meeting the radius $O'C$ at E.

In *Cases I and II*, complete the figure by drawing the arcs HK and KC, concentric with O and O' respectively.

Then ADF, GBE and HKC are the curves assumed by the horizons that pass through A, B and C respectively,

and $AH = DK = FC =$ the thickness of beds exposed between A and C.

In *Case III*, the centre O is cut out by the arc CH, and the thickness of beds exposed must be measured by FC.

It will be noticed again that all the arcs are tangential to one another, and the resulting curves are perfectly smooth. There is also a change in all three cases from the anticlinal sense to the synclinal sense or vice versa, as we proceed from left to right of the section.

Prop. 5. Given any four dip readings in the same direction, construct the curve assumed by the strata, and measure the thickness of beds exposed between the two outer dip readings.

The construction is the same as that of the two previous propositions, but we will take a case where the normals to the two inner dip readings intersect near ground surface, and show how we may still measure the thickness required.

For let A, B, C and D be the given dips.

Draw the normals, and construct those curves which are controlled by the four dips respectively, viz. the curves $EBLF$, $GKCH$, $RQPD$.

Inspection will show us that in the curve from the outcrop at A, the centre O' is cut out.

From F, along $O''F$ produced, mark off FM equal to EA.

With centre O and radius OA, draw the arc AN, and with centre O'' and radius $O''M$, draw the arc MN, intersecting the arc AN at N.

Then ANM is the curve required.

It will be noted that there is an angle at N, and in nature this is of course to some extent smoothed off. How much it is shown smoothed off in the section depends upon

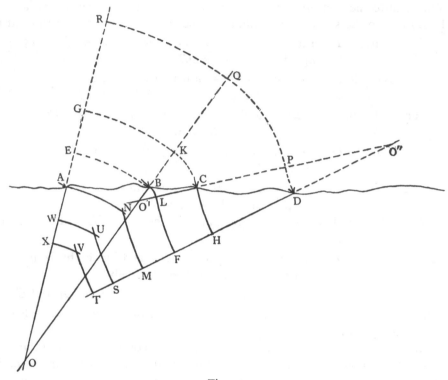

Fig. 19.

the scale, for, if denudation be carried to such a depth as to expose this apical angle, and the scale is increased, other dips round the apex will be recorded, which would give a smooth curve. It will be noted that:

$$AR = MD = \text{thickness of beds exposed between } A \text{ and } D.$$

Proceeding further,

on $O''T$, take the points of intersection of any other two horizons below M, e.g. S and T.

Make $AW = MS$, and $AX = MT$.

With centres O and O'' respectively, draw the arcs

$$WU, XV, \text{ and } TV, SU,$$

intersecting at U and V respectively.

The intersecting arcs WUS and XVT represent the curve of the two horizons from S and T respectively.

On what curve do the points O', N, U, V lie? We shall deal with this later, but we may say here that the curve is a conic.

We have considered so far a series of dips in one direction, but the same methods of construction may be applied to opposed or concurrent dips, until a complete anticline or a complete syncline, or a series of both is the result. Let us consider a possible instance, construct the actual figure and inspect it.

Prop. **6. Given any number of dips at outcrop, construct the curve for a certain given horizon, and others at unit thickness above and below it.**

Let dips be observed at the points
$A, B, C, D, E, F, G, H, I.$

Draw the successive normals, which intersect at
$O, O^1, O^2, O^3, O^4, O^5, O^6, O^7$ respectively.

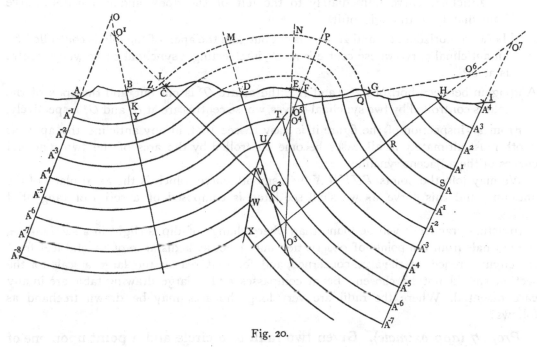

Fig. 20.

Let the given horizon, whose curve is required, and is such that it is controlled by arcs drawn from centres O, O^1, O^2, etc., and does not cut out any of these centres, outcrop at A.

With centre O and OA as radius, draw the arc AK,
which cuts OB produced at K.

With centre O^1 and radius O^1K, draw the arc KL,
which cuts O^1C at L.

Similarly describe the other arcs, LM, MN, NP, PQ, QR, RS.

Then $AKLMNPQRS$ is the curve required.

Let A^{-1} be a point on OA unit thickness below A.

From K along O^1K, mark off this unit thickness, KY, and proceed to mark off unit thicknesses from L, M, N, etc., along LO^2, MO^3, NO^4, etc., respectively.

Now with respective centres O, O^1, O^2, etc., describe the successive arcs at unit thickness from those which compose the curve $AKLMNPQRS$.

Proceed similarly for horizons at A^{-2}, A^{-3}, A^{-4} and for horizons above the curve $AK...S$, A^1, A^2, A^3, etc.

Now from the figure we note that centres O^4 and O^5 are cut out between horizons A^{-3} and A^{-4}, and that at A^{-4} and below that horizon the apex of the curve becomes an angle.

Thus A^{-4} is controlled in the crestal region by centres O^3 and O^6, and it is consequently an anticlinal curve immediately to the left of the apex and a synclinal curve immediately to the right of it.

And between horizons A^{-5} and A^{-6}, O^2 is cut out, and the apex of the fold is controlled by an anticlinal curve whose centre is at O^2, intersecting a synclinal curve whose centre is at O^6.

And again between horizons A^{-6} and A^{-7}, the centre O^3 is cut out, and the apex of the fold is controlled by two synclinal curves, whose centres are at O^1 and O^6 respectively.

From an inspection of the figure it is easy to see that, in any anticline, the apex at depth must ultimately in all cases become controlled by the arcs of the two synclinal curves of the adjacent syncline.

We may join the points $TUVWX$ by a smooth curve, which is the axial plane of the anticline, and this curve, as we shall see later, is composed of a series of tangential conics.

In actual practice it will be found that, where change of dip is slight in a section line, the normals from the points of observation may meet at a distance of several feet from the circumferences of the arcs concerned, and for that reason too large a scale for the section should not be chosen. Beam compasses and a large drawing table are in any case essential. Where the radii are very long the arcs may be drawn freehand as follows:

Prop. 7 (*approximate*). **Given two radii of a circle and a point upon one of them, which lies upon the circumference, draw approximately the arc of the circle that passes through the given point, without using the centre of the given circle.**

Let AB and CD be the two given radii and B the given point upon the radius AB.

From B it is required to draw the arc of the circle, whose centre is at the point of intersection of AB and CD produced.

Draw BC at rt. angles to AB.

Bisect BC at F.

Through F draw FG at rt. angles to CD, meeting CG at G.

Then *G* is the point of intersection of the arc with the radius *CD*, and the arc may be completed freehand from *B* to *G*.

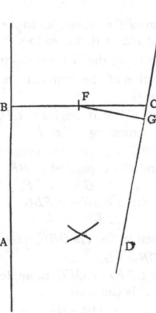

Fig. 21.

The mathematical fallacy here is, of course, that *FC* does not equal *FG*, but where the angle between *AB* and *CD* is small, the error is negligible.

Prop. 8. **Given two radii of a circle and a point upon one of them, which lies upon the circumference, draw the arc of the circle that passes through the given point, without using the centre of the given circle.**

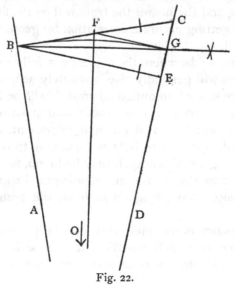

Fig. 22.

Let AB and CD be the two given radii and B the given point upon the radius AB, and upon the circumference of the given circle. Required to draw the arc of the circle from B to the radius CD.

Draw BC at rt. angles to AB,
and BE at rt. angles to CD.

Bisect the angle CBE, the bisector meeting CD at G.

Then G is the point of intersection of the required arc with CD, and the arc may be readily drawn freehand from B to G.

Proof.   Draw GF at rt. angles to CD,
meeting BC at F.

Now $\angle CBG = \angle EBG$ (by construction),
and FG is parallel to BE,
$\therefore \angle EBG = \angle BGF$,
and $\angle BGF = \angle FBG$,
$\therefore BF = FG$.

Then in the $\triangle$s OBF, OGF,

$$\because \begin{cases} BF = FG, \\ \angle OBF = \angle OFG \text{ rt. angle}, \\ FO \text{ is common}, \end{cases}$$

$$\therefore OB = OE$$

= the radii of the circle whose centre is at O.

The Interpolated Dip. Any section, drawn to scale on paper, must be strictly in accord with the evidence supplied by the geological map.

It would not do, of course, to adopt Prop. 6 and its construction without first proving that the horizon, which outcrops at A (Fig. 20), does actually crop out again at Z and again at G. The more abundant the evidence, the greater will be the number of arcs that it is possible to draw, and the nearer the truth will be the finished section; the more likely, too, shall we be to getting accurate correlation by geometrical construction direct from the evidence exposed.

In quite steeply-dipping folds, where the evidence is fully exposed, geometrical construction by tangential arcs will generally give absolutely accurate results; that is, when a horizon crops out on one side of an anticlinal crest, it will be brought to outcrop again by the series of arcs from observed dips in exactly that position that it occupies on the ground, thus proving that parallelism of horizons is present, and that the method of drawing is perfectly sound. But, where it is not possible to obtain a sufficiency of dip readings, and evidence is in general scanty, it may be found, for instance, in the anticline shown in Fig. 23, that, whereas the horizon outcropping at A should by construction from all the available dip evidence crop out again at B, on the ground it actually does so at (say) B' or B''.

Now, where this discrepancy is so small, that it is fairly evident that it is not caused by tectonic attenuation or by lateral variation, it will be legitimate to interpolate a reasonable dip, where the evidence is most scanty or doubtful, in such a manner as to bring about this adjustment, and when it is remembered that the number of dip readings which govern a given curve are in point of fact infinite, a small adjustment of this nature is quite reasonable.

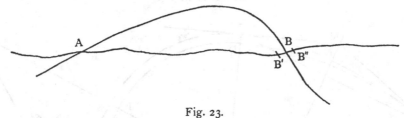

Fig. 23.

The construction for this interpolated dip is a somewhat complex matter, as it depends upon our finding two circles which are tangential to each other, and tangential to two given circles respectively, and the varieties of this proposition are very large. The construction for the interpolated dip, however, may, as we shall see, be used in cases where there is attenuation or lateral variation, and it is in point of fact a generally useful one ; it is consequently worth while examining the matter with some care. In the first place, there is a very simple construction, which, though only approximate, gives good results for small discrepancies, and also generally assigns a very natural position for the interpolated dip.

Prop. 9 (_approximate_). A curve constructed of circular arcs controlled by dip readings from one flank of a fold makes a near approach to a curve similarly constructed from the other, but does not meet it. It is required to interpolate a dip of such value and in such a position upon the one flank, that the resulting curve upon this flank will meet tangentially the curve upon the other.

As simple an instance as possible is taken.

Let A, B and C be the dips which control the circular arcs AD, DE and EH respectively, with respective centres O^1, O^2 and O^5.

By this curve the horizon, which crops out at A, should again crop out at F on the arc EH. But from actual observations in the field, it is found to crop out at G. Let it be reasonable to assume that a correcting dip occurs between A and B.

Through G, with centre O^5, draw the arc KG,

and with O^2 as centre and O^2K as radius, draw the arc LK, tangential to KG, and meeting the radius O^1D produced at L.

For the arc AL, it is required to substitute two arcs which are tangential to one another, and which are tangential to the arcs AD, LK at A and L respectively.

Draw AM at rt. angles to O^1A,

and LM at rt. angles to O^1L,

intersecting O^1A at M.

Join AL.

Through M draw the straight line MO^3O^4 at rt. angles to AL,

meeting O^1A at O^3,

and O^1L at O^4.

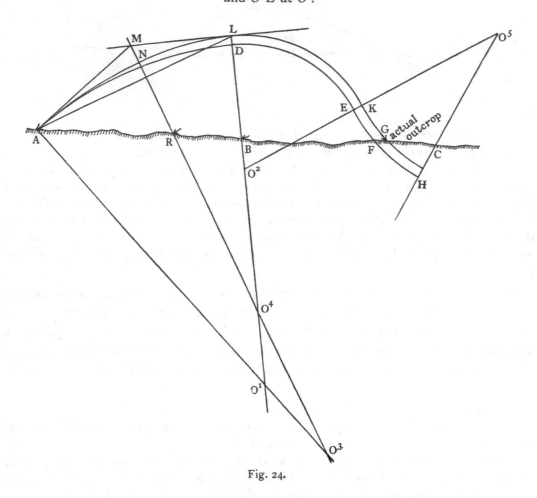

Fig. 24.

Then O^3 and O^4 are the centres of the two arcs required.

With centre O^3 and radius O^3A, draw the arc AN,

and with centre O^4 and radius O^4D, draw the arc ND.

Then if LD is small, AN and ND will approximately meet tangentially, and NO^4 is normal to the required dip, which may be interpolated at R.

There is no proof for this construction, which is approximate only, and can only be used for small values of DL.

Should DL be so large that AM and ML do not meet in such a position as to bring R between A and B, it is possible to repeat the construction for a second portion of the fold and interpolate two dips.

It will be seen that this approximate proposition definitely assigns a position and value for the interpolated dip R, and that it also gives a very reasonable curve. In point of fact, however, the possible positions for R and its value are infinite, though strictly within limits, and the radius of one of the arcs may be drawn arbitrarily, but within those limits, the radius and curvature of the second arc being dependent upon the choice of the radius and curvature of the first.

Accurate construction for the Interpolated Dip. To show this, and to obtain a construction for the interpolated dip, which is mathematically accurate, it is necessary to consider the following proposition*.

Prop. 10. Given any two circles of any radius, it is required to draw two other circles tangential to each other, one of which shall be tangential to one of the given circles, and the other tangential to the other.

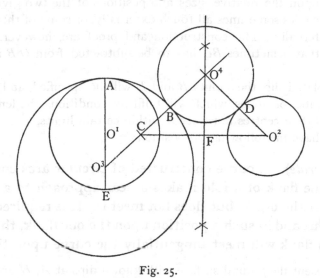

Fig. 25.

Let O^1 be the centre of the first given circle, and O^1A its radius.

Let O^2 be the centre of the second circle, and O^2D its radius.

Take any point A on the circumference of the first circle,
and, through A, draw the diameter AE.

Take any point O^3 on AE,
and with centre O^3, and O^3A as radius, describe a circle,
which will be tangential to the circle, whose centre is at O^1 at the point A.

Draw any radius O^3B and produce.

Along BO^3, cut off BC equal to O^2D.

---

* The author is indebted to Mr W. N. Bray for his valuable assistance in working out the following propositions for the interpolated dip.

Join O^2C and bisect it perpendicularly,
the bisector meeting O^3B produced at O^4.

With centre O^4, and with O^4B as radius, describe the fourth circle required.

Proof. The circle whose centre is at O^4 is tangential to the circle whose centre is at O^3 by construction.

And since the triangle O^4FC, O^4FO^2 are equal in all respects,
$$O^4C = O^4O^2.$$

But $BC = O^2D$ (by construction),
$$\therefore O^4B = O^4D.$$

$\therefore$ the circle whose centre is at O^4 is also tangential to the circle
whose centre is at O^2.

The variations of this proposition are many and depend upon the positions taken for A, B and D, also upon the relative sizes and positions of the two given circles. It will be found that the circles sometimes all touch externally, or some of them externally and some of them internally. The construction and proof are, however, the same in all instances, except that sometimes BC has to be subtracted from O^4B instead of added to it.

For the interpolated dip the points B and D will be specified, and though there are still an infinite number of circles, which will fulfil the conditions, the length of their radii and the position of their centres must occur within certain limits.

Proceeding, we have the following proposition:

Prop. 11 (*accurate*). **A curve constructed of circular arcs controlled by dip readings from one flank of a fold makes a near approach to a curve similarly constructed from the other, but does not meet it. It is required to interpolate a dip of such value and in such a position upon the one flank, that the resulting curve upon this flank will meet tangentially the curve upon the other.**

Let AHK represent the ground surface with known dips at A, H and K respectively.
Construct the normals, which meet at O^1 and O^2,
and draw the arcs AB and CK,
which meet the normal O^1O^2H produced, at B and C respectively,
in *Case I*, where B is on the side of C remote from O^1 and O^2,
and in *Case II*, where B is on the side of C nearer to O^1 and O^2.

For the arc AB, it is required to substitute two arcs, which are tangential to one another, and which are also tangential to the arcs AB and CK at A and C respectively.

The number of pairs of interdependent arcs that it is possible to substitute for the arc AB is infinite, but they all lie strictly within limits.

Limiting condition, No. 1, determining the relative positions of B, C and O^2.

In *Case I*, where B lies on the side of C remote from O^1 and O^2, the limiting condition is that the tangent at C must cut the arc AB. Draw AX perpendicular to BO^2. Then

C must lie on the side of X nearer to B. The points B and C may be separated by any distance within this limit, and O^1 and O^2 are interchangeable.

In *Case II*, where B lies on the side of C nearer to O^1 and O^2, draw the tangent to the arc AB at A, meeting BC produced at X. Then C must lie between B and X.

Limiting condition, No. 2, determining the position of O^3, the centre of the first of the two circles required.

In *Case I*, bisect AC perpendicularly, the bisector meeting O^1O^2 at Z. Then O^3 must lie on O^1O^2 produced on the side of Z remote from O^1 and O^2, thus assuring that the arc from C with radius O^3C cuts the arc AB.

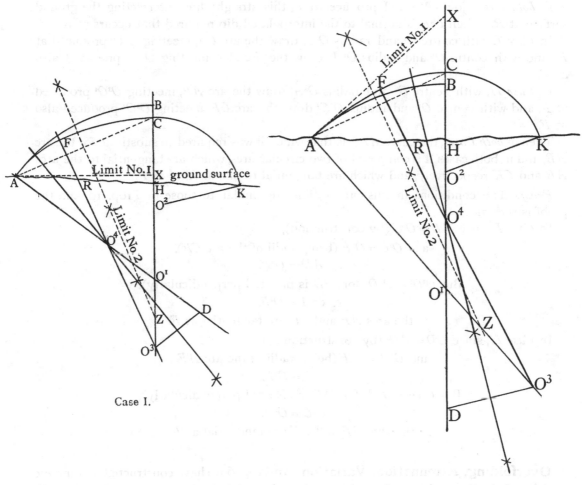

Case I.

Fig. 26.

In *Case II*, bisect AC perpendicularly, the bisector meeting O^1A at Z. Then O^3 must lie on AO^1 produced on the side of Z remote from O^1, thus assuring that the arc from A with radius O^3A cuts XC.

Continuing the proposition within these limits, choose O^3.

In *Case I*, from A along AO^1 produced, mark off AD equal to O^3C, and in *Case II*, from C along CO^1 produced, mark off CD equal to AO^3.

In *both Cases*, join O^3D, and bisect it at rt. angles.

In *Case I*, the bisector meeting AD at O^4, and in *Case II*, the bisector meeting CD at O^4.

Then O^4 is the centre of the fourth circle required.

In *both Cases*, join O^3O^4 and produce to F, this straight line intersecting the ground surface at R. Then O^4R is normal to the interpolated dip required that occurs at R.

In *Case I*, with centre O^3 and radius O^3C, draw the arc CF, meeting O^3O^4 produced at F, and with centre O^4 and radius O^4A, draw the arc AF, meeting O^3O^4 produced also at F.

In *Case II*, with centre O^3 and radius O^3A, draw the arc AF, meeting O^3O^4 produced at F, and with centre O^4 and radius O^4C, draw the arc CF, meeting O^3O^4 produced also at F.

Then in *both Cases*, the curve AFC is that which it was required to substitute for the 'arc AB, and in both cases it is composed of two circular arcs, which are tangential to the arcs AB and CK respectively, and which are tangential to one another at F.

Proof. The conditions in this proposition are similar to those in Prop. 10, and the proof is the same.

In *Case I*, since $AD = O^3C$ (by construction),

and $O^3C = O^3F$ (being radii of the arc CF),

$$\therefore AD = O^3F.$$

But $O^3O^4 = O^4D$ (for O^3D is bisected perpendicularly),

$$\therefore O^4A = O^4F,$$

$\therefore$ the arcs AF and FC are tangential at F.

In *Case II*, since $CD = AO^3$ (by construction),

and $O^3A = O^3F$ (being radii of the arc AF),

$$\therefore CD = O^3F.$$

But $O^3O^4 = O^4D$ (for O^3D is bisected perpendicularly),

$$\therefore O^4C = O^4F,$$

$\therefore$ the arcs AF and FC are tangential at F.

Overfolding, Attenuation, Variation. In Prop. 6 we have constructed a complete anticline from the evidence afforded by surface dips. We have now to consider how far surface evidence will enable us to construct an anticline with an overfolded limb, and also to consider what overfolding implies.

Prop. 12.   **In any set of competent overfolds, the condition that there should be surface evidence showing a smooth curve both in an anticline and in its adjacent foresyncline * (i.e. when the middle limb between them is reversed), along any specified horizontal plane, is impossible.**

Take a simple instance:

Let *ABCDE* represent the curve of a horizon in any anticline and its adjacent

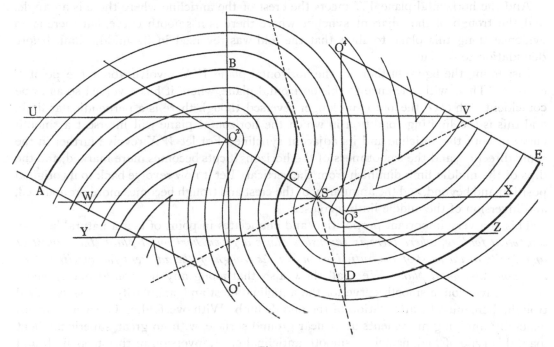

Fig. 27.

foresyncline, the middle limb *BCD* being overfolded. Let the curve of this horizon be composed:

> of the arc *AB*, whose centre is at O^1,
>> tangential to the arc *BC*, whose centre is at O^2,
>> tangential to the arc *CD*, whose centre is at O^3,
>> tangential to the arc *DE*, whose centre is at O^4.

Complete the figure for horizons unit thickness apart.

Let the horizontal plane be parallel to *UV*, *WX* and *YZ*.

Then from an inspection of the figure it is clear that no straight line parallel to the horizontal plane can be drawn so that it cuts across the middle limb, where it is inverted, and across a smooth curve either in the anticline or in its foresyncline.

---

* The term "foresyncline" is used throughout this work to indicate the syncline adjacent to an asymmetric anticline upon its steep side. The "middle limb" is the steep limb between an asymmetric anticline and its foresyncline.

For the horizontal plane UV cuts across the smooth curve of the anticline, and meets the syncline where there is an angle in the trough, but there is no evidence here that the middle limb is overfolded at depth, though such might be suspected from the acuteness of the trough of the syncline.

And again, the horizontal plane WX meets both the anticline and its adjacent syncline, where both crest and axis are represented by an angle, and the middle limb is overfolded along that line.

And the horizontal plane YZ meets the crest of the anticline where there is an angle, and the trough of the adjacent syncline where there is a smooth curve, but there is no evidence along this plane to show that the fold was reversed in its middle limb before denudation took place.

Inspecting the figure further, let the horizontal plane WX revolve about the point S on O^2O^8. Then, with reference to this horizontal plane, where it has revolved so as to be coincident with O^2O^8, the fold is no longer reversed, though there is a perpendicular limb, and this is the limiting condition in which the horizontal plane will intersect a smooth curve in both the anticline and its adjacent syncline. For let WX revolve further in the same direction, and the smoothness of the horizons it cuts becomes more marked, and the dips of the middle limb diminish from the vertical. Let it now revolve back to its original position and beyond, and the angularity of the crest and trough becomes more pronounced, and inversion of the middle limb increases.

This is a very important proposition, and enunciates in point of fact a natural law. *In competent folding, where there is inversion along a given horizontal plane, there must be angularity at the crest of the anticline, and in the trough of the adjacent foresyncline along the same horizontal plane.* We shall show also that along any horizontal plane, where there is inversion, a smooth curve for the anticlinal crest and angularity of the synclinal trough, there must be attenuation of the middle limb. With overfolding in Tertiary rocks, generally implying movements at or near ground surface, with no great superincumbent load, this type of fold, namely a smooth anticlinal crest, inversion in the steep limb and an angular syncline or "synclinal bend*," all occurring in the same horizontal plane, is very common, and the mapping of the synclinal bend and determination of the magnitude of the attenuation in such instances are of the first importance.

Prop. 13. In any fold controlled by dips along a horizontal plane, which give a smooth anticlinal crest and inversion of the middle limb, there must be incompetency and attenuation of that limb.

Let horizon $AKLD$ be controlled by the dips A, B, C and D along the horizontal plane AD, and let there be a reversed dip at H.

From A^1, unit thickness above horizon $AKLD$, construct the competent curve A^1F.

From F it is only possible to maintain the thickness AA^1 by drawing the arc FG,

with D as centre and DF as radius, thus giving verticality at G.

But the beds at G are not vertical, but reversed.

---

* For previous use of this term cf. Pascoe, *The Oilfields of Burmah.*

Draw HO^4, normal to the dip at H,
and with centre O^4 and radius O^4F, draw the arc FK.
Then FK is the arc which accords with the evidence of inversion at H.
But DK is less than DG.
∴ there is attenuation of the middle limb.

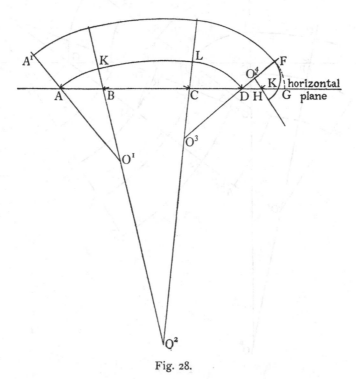

Fig. 28.

In this proposition it can be seen that the introduction of the centre at O^4, with the radius O^4H, which passes below the horizontal plane, leads us immediately into uncertainty in regard to the completion of the section line, however simple and straightforward the other evidence upon the steep flank beyond K may be. For to bring the beds to surface again, a centre below the horizontal plane on O^4H produced must be chosen, and for this centre there can be no direct evidence along the horizontal plane. By means of either of the propositions for the interpolated dip, arcs may be chosen for and inserted below the horizontal plane, but, where there are a number of horizons, none of which remain parallel, the construction becomes so complicated that a resort to freehand drawing is generally justified.

Repeating the evidence exposed for the next proposition, and adding other items that may be known upon the steep flank, and keeping the example in its simplest form, we have the following:

Prop. 14. **Given a fold controlled by dips along a horizontal plane, competent upon one flank, but inverted and attenuated upon the other, construct the fold so far as the evidence will allow.**

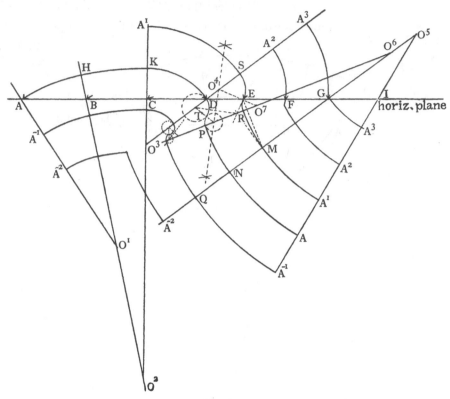

Fig. 29.

Let the flexure $AHKD$ be controlled by dips at A, B, C and D, the horizon A cropping out again at D. Let horizon A^1, unit thickness above horizon A, be found to be inverted at E, and let the position of the synclinal bend occur at G, where horizon A^3 meets the horizontal plane.

The centres controlling the competent part of the fold are

$$O^1, O^2, O^3, O^4 \text{ and } O^5.$$

Complete the figure only in so far as it remains competent.

Considering horizon A^1, we find that this may be drawn competently as far as E on the one side, and M on the other.

By Prop. 9, we may now interpolate the arcs MR, RE, which are tangential to one another, and tangential to the arcs MA^1 and ES respectively.

The two new centres found are O^6 and O^7.

The point at which O^6R meets the horizontal plane does not of course imply a dip normal to it in that position.

Considering horizon A, we may now carry this competently to P, by using the centre O^2; and we have the arc KD on the other side.

By Prop. 10, we now interpolate the arcs DT and TP.

Similarly with horizon A^{-1}.

At horizon A^{-2}, the centre O^3 cuts out, and the fold becomes competent.

Inspecting the figure further, we note that the number of varieties for the small arcs DT, TP, etc., is infinite, and bearing in mind the fact that for each and every horizon drawn they will be different, it is doubtful whether there is much advantage in using the added complication of geometrical construction for them. In all cases of overfolding, all we can do is to construct the fold competently, as far as it remains competent, and from that point onwards to consider each case upon the merits of its own evidence. There is bound to be certain guesswork, and the evidence along a single horizontal plane can never give a complete solution. Our work should be directed to reducing this guesswork to a minimum.

Very much the same argument applies to *lateral variation*. Where thickening or thinning occurs along a section line at a definite rate, the construction for the interpolated dip may be used to connect horizons across from radius to radius, but again a different construction will be required for each horizon.

CHAPTER IV

THE AXIAL PLANE

WE have shown by our constructions in the previous chapter that arcs in the synclinal sense expand at depth, and that arcs in the anticlinal sense contract and cut out. Thus the axial plane of an anticline (see Definition 1, p. 7) is ultimately controlled by two arcs far out upon the flank.

For the purposes of oil finding a correct determination of the axial plane is one of the first essentials, and much money has been wasted by oil companies through the failure of their geologists to grasp the significance of the controlling factors which determine its behaviour. In anticlines that are even, in a moderate degree only, asymmetric, it is never safe for the drill to cross the axial plane, and to penetrate into the steep flank, and, where a test well has been proved to have done so, its evidence in regard to the oil prospects of the area can generally be ruled out as inconclusive. It should be noted in the first instance that the "axial plane" is not a plane at all, but a highly complex curved surface, though the term "plane" is now so generally used that it is here retained. It is, as we shall show in competent folding, a surface which may be defined by a number of tangential conics.

The first requirement for the determination of the axial plane is the representation of the fold in section correctly[*]. We have shown that for competent folding this can be done with mathematical accuracy, but that, where there is attenuation or variation, there is bound to be some uncertainty. Nevertheless, the behaviour of the axial plane in competent folding has a very wide bearing upon its behaviour, where there is attenuation and incompetency, and a thorough knowledge of the determinable curvature it may adopt in competent folding is of great importance. Furthermore, very many folds are, in point of fact, strictly competent, and it is in dealing with these that as many errors have been made in the past as with the more complex question involving attenuation of the middle limb and overthrusting.

In the survey of a promising structure it has been very common to omit the greater part of the very necessary work upon the steep flank of the anticline on the principle that it is not here, at any rate, that the oil field lies; and, though of course this is so, it is the steep flank of the fold that is the most important factor in the determination of the axial plane, and, as can be seen from Figs. 30 and 31, the dips which control the axial plane at a given depth are relatively further out from the crest upon the steep flank than are the controlling dips for the gentle flank.

Turning to Fig. 30, which represents an asymmetric competent fold, i.e. a fold with no attenuation of the middle limb, controlled by arcs, whose centres are at O^1, O^2, O^3, O^4, respectively, we may note that it has been common practice to carry mapping in the field only as far as (say) the point A, and to assume that the great thickness of beds occurring with steep dips is repeated outwards and to infinity. The dotted lines in the figure

[*] Cf. E. H. Cunningham Craig, *Oil Finding*, 1920, pp. 237 *et seq.*

illustrate the danger of such an assumption; for a well drilled to strike at depth the crest
of the horizon *CDEF* would only meet it at some distance down the flank, and, as can

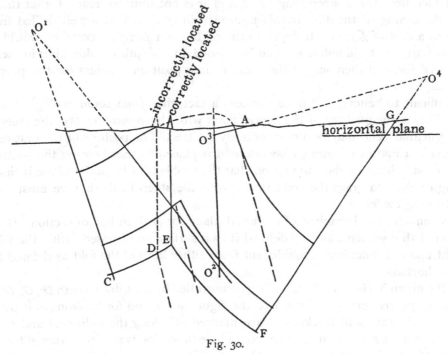

Fig. 30.

be seen, the axial plane at this depth is controlled by the dip *G*, far out upon the steep
flank.

A still more dangerous assumption, but of a similar nature, is shown in Fig. 31, where

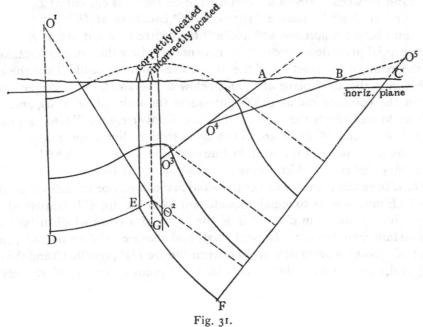

Fig. 31.

in a broad flat-topped anticline, the dip A is taken as generally representative of the steep flank, and the fold constructed accordingly, as is shown by the dotted lines. There is, in point of fact, however, a steepening at B, and it is not until we reach C that there is a general slackening of the dip. In this figure it will be seen that a well drilled in error, possibly at a cost of £30,000, if depth to the oil horizon DEF is 3000 feet, would fail to meet that horizon at all, unless it could be carried to a depth considerably in excess of that allowed for, and then only at the point G, in a position hopeless for the purpose of finding oil.

It is difficult to believe, still none the less a fact, that from some geological surveys, which delineate the geology of the gentle flank with the utmost nicety, the steep flank and its synclinal bend may be almost entirely omitted. The position of the first test well is then made a matter of pure guesswork, and is placed "somewhere on the gentle flank near the crest" either on the assumption that the problem of the axial plane is insoluble, or through failure to grasp the point that it is to the steep flank that we must look for the controlling evidence.

Let us consider the behaviour of the axial plane for a given line of section. It will be remembered that we have already defined it as a surface so disposed within the fold that any point upon that surface is equidistant from either limb of the fold as defined by any particular horizon.

Take the given horizon α, whose folding is controlled by arcs, drawn with O^1, O^2, O^3, O^4, O^5 and O^6 as respective centres. Complete the figure for α, and for horizons unit thickness below α. To do this, unit thicknesses are marked off along the radii O^1A and O^6B, and the successive arcs are drawn in. The axial plane in section is now constructed by joining the apices of each horizon with a smooth curve. It will be noted that centres O^4, O^3, O^2 and O^5 cut out in succession.

Now between horizons α^{-3} and α^{-4} the centre O^4 is cut out at O^4,

and between horizons α^{-4} and α^{-5} the centre O^3 is cut out at C,

and at about horizons α^{-7} the centre O^2 is cut out at D,

and between horizons α^{-12} and α^{-13} the centre O^5 is cut out at E.

Tracing the axial plane down from O^4, we may notice where this cutting out takes place, namely at the point of intersection of the successive common radii. Thus the common radius O^2O^3U cuts the axial plane at C, and below C the arc UV no longer affects the curve. Again the common radius O^1TO^2 intersects the axial plane at D, and below D the arc TU no longer affects the curve, and so on. Conversely it will later be proved that the points O^2, O^3, C, and O^2, D, O^1 are in the same straight line respectively.

Above O^4 the axial plane is the straight line bisecting the angle VO^4W.

Now, analysing the curve, which forms the axial plane, we have:

From O^4 to C the axial plane (or rather the axial plane in section) is defined by the locus of a point which moves so as to remain equidistant from the arc UV (anticlinal) and the arc WX (synclinal); and from C to D, it is the locus of a point which moves so as to remain equidistant from the arc TU (anticlinal) and the arc WX (synclinal); and from D to E, that of a point moving similarly between the arc UV (synclinal) and the arc WX (synclinal); and, finally, from E downwards, that of a point moving similarly between the arc UV (synclinal) and the arc XY (synclinal). From O^4 upwards, it is the locus of a point moving equidistantly from the arc UV and the arc WX, and is consequently the

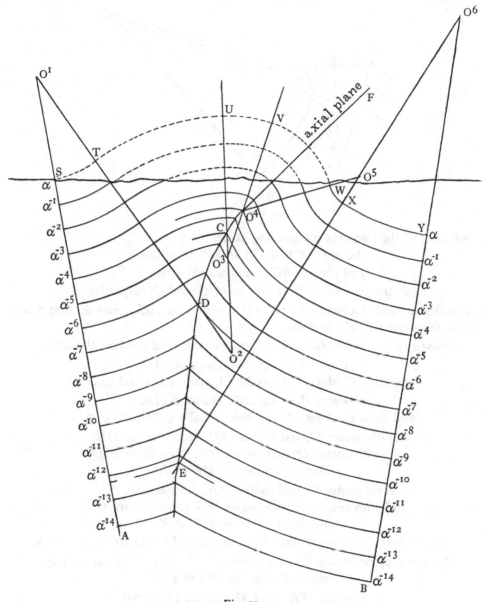

Fig. 32.

bisector of the angle VO^4W, i.e. a straight line. Elsewhere it is a series of curves, controlled by the above conditions. What are these curves, and how are they related?

Let us take each of the possible combinations of intersecting anticlinal and synclinal arcs, and discover what is the locus of a point which moves equidistantly between them.

Prop. 15. To find the axial plane * between two intersecting arcs, the one being in a synclinal sense, the other in an anticlinal sense.

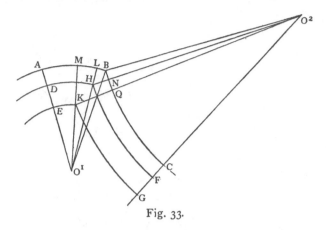

Fig. 33.

Let AB and BC be two arcs intersecting at B,

the arc AB being in an anticlinal sense,

and the arc BC in a synclinal sense.

Let the centres of these arcs be at O^1 and O^2 respectively.

It is required to find the locus of a point that moves in such a manner that it is always equidistant from the arcs AB and BC.

Draw any two radii O^1A and O^2C respectively and produce the latter.

From A, along AO^1, mark off AD,

and from C, along O^2C produced, make CF equal to AD.

From A, along AO^1, mark off AE,

and from C, along O^2C produced, make CG equal to AE.

With centre O^1, and radius O^1D, draw the arc DH,

and with centre O^2, and radius O^2F, draw the arc FH,

intersecting the arc DH at H.

With centre O^1, and radius O^1E, draw the arc EK,

and with centre O^2, and radius O^2G, draw the arc GK,

intersecting the arc EK at K.

Then H and K are equidistant from the arcs AB and BC respectively.

Now it may be shown that the points B, H and K lie on an ellipse,

whose foci are at O^1 and O^2.

For join O^1K, O^1H, O^1B, O^2K, O^2H, O^2B.

Since the sum of the focal distances of any point on an ellipse is constant,

it is required to prove that $O^1K + O^2K = O^1H + O^2H = O^1B + O^2B$.

* In the following propositions the term "axial plane" is used to denote the axial plane in geological section. The line of the horizon is considered to be parallel with the top and bottom edge of the page.

Produce O^1K to meet the arc AB at M, and
produce O^1H to meet the arc AB at L,
and let O^2H and O^2K intersect the arc BC at N and Q respectively.
Then $HL = HN$, and $KM = KQ$,
also $O^1M = O^1L = O^1B$ (radii),
and $O^2Q = O^2N = O^2B$ (also radii).
$$\therefore\ O^2Q + QK + O^1K = O^2N + NH + HO^1 = O^2B + O^1B,$$
the sum of the radii of the two arcs.

$\therefore$ the points B, H and K lie upon an ellipse, whose foci are O^1 and O^2.

Similarly with any other points, which lie equidistantly from the arcs AB and BC.

Therefore the axial plane in section between the two arcs
AB and BC is an ellipse with O^1 and O^2 as foci.

The ellipse may be drawn in the usual way with string stretched to the right length by the point of a pencil between O^1 and O^2.

Examining this proposition a little further, we may complete the arcs (Fig. 34), and we may note that the ellipse passes through the points of intersection of the two circles, or that the sum of the focal distances of any point on the ellipse is equal to the sum of the radii of the two circles.

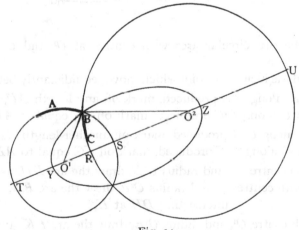

Fig. 34.

Also, joining the two foci and producing in either direction, we have:
$$O^1Y + O^2Y = O^1B + O^2B = O^1S + O^2R = O^1R + O^2S + 2RS.$$
$$\therefore\ 2O^1Y + RS = 2RS,$$
$$\therefore\ 2O^1Y = RS,$$
and $YT = YR$, $ZS = ZU$.

Every point on the ellipse is equidistant from the two circumferences.

It will be obvious from the figure that there are limits in nature to the length of the arcs AB and BC, and neither circle can ever be so much as half completed, so that a complete ellipse for an axial plane can never materialise. If the arc AB is carried down to where it assumes verticality, it is clear that the axial plane will also become vertical.

Reversing the figure we may note that we again have two arcs, this time forming a syncline, the one arc in an anticlinal sense, and the other in a synclinal sense. The axial plane is, as before, the same ellipse.

Prop. 16. **To find the axial plane between two intersecting arcs, both being in a synclinal sense.**

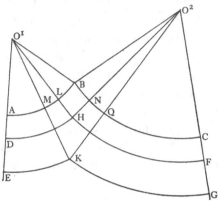

Fig. 35.

Let AB and BC be two circular arcs with centres at O^1 and O^2 respectively, and intersecting at B.

It is required to find the locus of a point which moves equidistantly between AB and BC.

From A, along O^1A produced, mark off any length AD, and
from C, along O^2C produced, mark off CF equal to AD.

From A, along O^1A produced, mark off another length AE, and
from C, along O^2C produced, mark off CG equal to AE.

With centre O^1, and radius O^1D, draw the arc DH, and
with centre O^2, and radius O^2F, draw the arc FH,
intersecting DH at H.

With centre O^1, and radius O^1E, draw the arc EK, and
with centre O^2, and radius O^2G, draw the arc GK,
intersecting EK at K.

Then B, H and K lie equidistantly from the arcs AB and BC, and it may be proved that B, H and K lie on a hyperbola, whose focal points are at O^1 and O^2.

For join O^1K, O^1H intersecting the arc AB at M and L respectively.
And join O^2K, O^2H intersecting the arc BC at N and Q respectively.

Join also O^1B, O^2B.

Now a hyperbola is the locus of a point, which moves in a plane containing a given point called the focus, and a given straight line called the directrix, in such a manner that its distance from the focus is in constant ratio greater than unity to its perpendicular distance from the directrix.

And from this it may be proved that*:

The difference of the focal distances of any point on a hyperbola is constant.

Now from the figure:

$O^2B - O^1B =$ the difference between the radii of the arcs BC and AB respectively,

and $LH = NH$ (by construction).

$$\therefore O^2H - O^1H = O^2B - O^1B.$$

Similarly $O^2K - O^1K = O^2B - O^1B.$

B, H and K lie on a hyperbola, whose focal points are at O^1 and O^2.

We have noted that in any anticline the axial plane will always be ultimately controlled by two intersecting arcs in the synclinal sense. The intersection of two synclinal arcs is therefore the most important combination with which we have to deal. We must consequently consider the properties of the hyperbola which forms the axial plane between them in some detail.

It is necessary first, however, to define some of the terms which are used in connection with this curve, and describe some of its relationships. In the first place, unlike an ellipse or a parabola, the hyperbola has two separate branches, though, in point of fact, in our section drawing we shall only be dealing with one of them. Again, from the fact that the

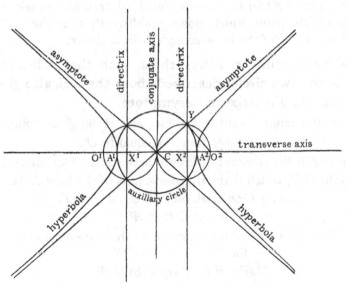

Fig. 36. Illustrating the terminology and relationships
of the hyperbola.

difference of the focal distances of any point on a hyperbola is constant, if we are given any one point upon a hyperbola, and its two foci, we may construct the whole curve and find its directrices. In such a manner Fig. 36 was constructed. Certain properties follow

* Charles Smith, *Geometrical Conics*, 1926, p. 124.

from the relationship of the curve to its foci and directrices, and these will be found proved in any text-book on geometrical conics.

Inspecting Fig. 36, we may observe:

(1) The *transverse* and *conjugate axes* of the curve, about both of which it is symmetrical.

(2) The *centre* of the curve (C), at the intersection of the transverse and conjugate axes, and consequently midway between the foci O^1 and O^2.

(3) The *vertices* of the curve A^1 and A^2 on the transverse axis.

(4) The *auxiliary circle* described with C as centre and CA^1 as radius, and consequently touching the two branches of the curve, where they intersect the transverse axes, at their vertices.

(5) The *directrices*, which may be proved to cut the auxiliary circle at Y, where CYO^2 is a right angle. Hence the directrices may be found by describing semicircles on CO^1 and CO^2 as diameters, and joining the appropriate points of intersection with the auxiliary circle*.

(6) The *asymptotes*, which are tangents to the curve, where the point of contact is at an infinite distance. The asymptotes intersect at the centre, C.

The asymptotes of the hyperbola are of great importance to us, for it will be shown that the axial plane in any competent anticline is ultimately coincident with the asymptotes of that hyperbola, which is controlled by the two ultimate synclinal arcs.

Let us now return to our two intersecting synclinal arcs, and consider the hyperbola traced by the locus of the point, which moves equidistantly from them. Let us complete the figure so that the whole of the intersecting circles is shown.

Prop. 17. Given a point on a hyperbola, such that it lies at the point of intersection between two circles described about the foci, also given, construct the curve and find its directrix and asymptote.

Let O^1 and O^2 be the required centres and foci, and P the given point.

Along PO^2, draw PT equal to O^1P.

Then O^2T is the constant difference between the focal distances.

Join O^1O^2, which is the transverse axis of the hyperbola.

Along O^2O^1 mark off O^2R equal to O^2T.

Bisect O^1R at A^1.

Then A^1 is a point on the curve on the transverse axis,
for $O^2A^1 - O^1A^1 = O^2R$.

Mark off O^2A^2 equal to O^1A^1.

Then A^2 is a point on the other branch of the curve on the transverse axis.

Bisect A^1A^2 at C,

which is the centre of the hyperbola and of the auxiliary circle.

With C as centre, and radius CA^1, draw the auxiliary circle.

On O^1C as diameter describe a circle cutting the auxiliary circle at Y and Y'.

* Charles Smith, *op. cit.* p. 136.

Then CY and CY' produced are the asymptotes of the hyperbola, and YY' one of the directrices.

Other points on the hyperbola may be found and the curve thus completed by taking any radii of the respective intersecting circles and producing them by equal lengths, as is shown in the figure; thus with O^1 and O^2 as centres respectively, and O^1H and O^2K as radii, describe two arcs intersecting at M. Then M is a point on the hyperbola.

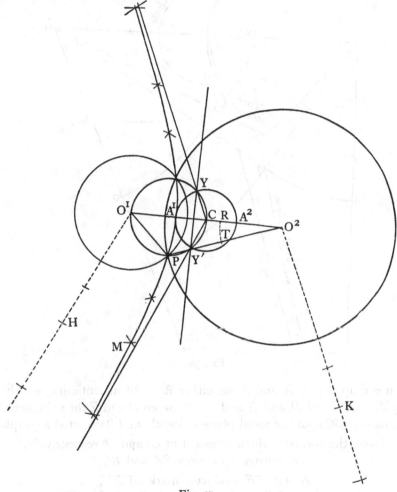

Fig. 37.

We are now in a position to take any anticlinal fold and determine the ultimate position of its axial plane, even though the evidence along the crest of the anticline is obscured. We will take here a simple case only, in which the flanks of the fold are synclinal arcs of circular section. This construction is useful also where there is no time for a complete survey of a fold, but where the outcrop of a known horizon on either flank has been determined.

6-2

Prop. 18. Given an anticline, whose flanks are controlled by two synclinal arcs of circular section, and given a known horizon which crops out on either flank. Find the ultimate position of the axial plane at depth, and the related asymptote of that hyperbola, by which it is formed.

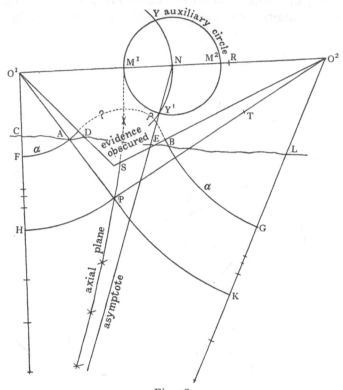

Fig. 38.

Let horizon α crop out at A and B on either flank of an anticline, which flanks are controlled by dips at C and D, and E and L. Between D and E the evidence is obscured. Find the ultimate position of the axial plane at depth, and its related asymptote.

Draw the normals, which intersect at O^1 and O^2 respectively,
and construct the arcs FA and BG.

Along O^1F produced, mark off FH,
and along O^2G, mark off K, such that $FH = GK$.

With O^1 as centre, and radius O^1H, construct the arc HP, and
with O^2 as centre, and radius O^2K, construct the arc KP,
intersecting the arc HP at P.

N.B. The distance FH must be of such magnitude as to ensure the intersection of the arcs.

Then P is a point on the hyperbolic axial plane.

Join O^1O^2, O^1P, O^2P;

and along PO^2, mark off PT equal to PO^1.

Then O^2T is the constant difference between the focal distances of the hyperbola.

Along O^2O^1, mark off O^2R equal to O^2T.

Bisect O^1R at M^1.

Then M^1 is one of the vertices of the hyperbola.

Mark off O^2M^2 equal to O^1M^1.

Bisect M^1M^2 at N.

And with centre N, and radius NM^1, draw the auxiliary circle.

On O^1N as diameter describe a circle cutting the auxiliary circle at Y and Y'.

Then NY' is the asymptote required.

The hyperbola may be drawn as shown in Prop. 17, and the axial plane may be carried up to the point S.

From this proposition it will be noted that the axial plane can never cross the asymptote, and wells to strike the axial plane at depth can be safely located without reference to the crestal part of the fold.

It will also be noted that general absence of evidence near the crest of the fold may be immaterial, whereas evidence from some distance down the flank is of the utmost importance. When in a newly discovered anticline, which it is desired to test for oil, there is abundant evidence on the crest, but little or nothing on the flanks, it may be desirable to dig shafts or trenches in order to procure the flank dips for the determination of the axial plane. It is most important to recognise that in nearly all instances it is the flank dips that are required rather than those around the crest, and the cost of an inconclusive or unsound test well, whose position rests upon the evidence of the latter, many times outweighs the small expense of sinking a shallow shaft in order to procure the former, which alone can determine the correct location.

Having considered the axial plane between:

   (1) A synclinal and an anticlinal arc; and

   (2) Two synclinal arcs;

and shown that in (1) it is an ellipse, and in (2) it is a hyperbola, we have, in point of fact, dealt with all the possible variations. For the case of two anticlinal arcs intersecting is merely the reverse of the case of the intersection of two synclinal arcs, and this may be observed from Fig. 39, where the various cases are shown.

For let O^1 and O^2 be the centres of two circles intersecting at P and H.

Then KH and HL are two arcs intersecting in a synclinal sense,
and the axis between them is a hyperbola.

But MP and PB are two arcs which intersect, the one in an anticlinal sense,
and the other in a synclinal sense, and the axis between them is an ellipse.

The arcs BP and EP are in reality both in an anticlinal sense, though in nature so acute an angle between them would be very unusual, or even impossible. The more usual

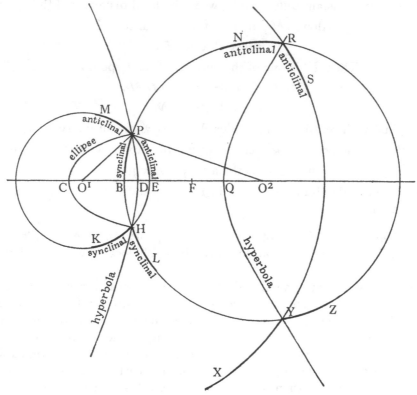

Fig. 39. Illustrating the combination of various forms of arcs, and their relative axial planes.

form for the intersection of two anticlinal arcs is where the radius of one of the circles is greater than O^1O^2, as in the case of the arc RS, whose centre is at O^1, and the arc NR, whose centre is at O^2. The axis will again be a hyperbola, but we will consider this case in another proposition. The arcs XY and YZ, both in a synclinal sense, will also be noted, though their intersection at Y is very obtuse.

Before leaving the figure it is worth while looking at some further relationships. For considering the ellipse PCH,

$$2O^1C = BE \text{ (Prop. 15; Fig. 34)}.$$

And considering the hyperbola PDH,

on the transverse axis O^1O^2 mark off $O^2F = O^2P - O^1P$.

$$\text{Then } O^2D = O^2B - BD,$$
$$\text{and } O^1D = O^1E - DE,$$
$$\text{and } O^2D - O^1D = O^2F.$$

$$\therefore\ O^2B - BD - O^1E + DE = O^2F;$$

$$\text{also } O^1E = O^1P,$$

$$\text{and } O^2B = O^2P.$$

$$\therefore\ O^2P - O^1P + DE = O^2F + BD.$$

$$\therefore\ BD = DE.$$

It is interesting to note that the ellipse and the two hyperbolas shown are confocal conics, and as such cut one another at right angles at their common points.

We may also consider certain limiting conditions for the hyperbolic axial plane formed by the intersection of two arcs whose curvature is in the same sense, either synclinal or anticlinal. For let the two foci, which are also the centres of the respective arcs, and the radius of curvature of one of the arcs remain fixed, and let the radius of curvature of the second vary.

Then it is clear that (Fig. 40):

(1) When the two circles are equal, i.e. when the anticline is symmetrical as far as its flank dips are concerned, the curve is a straight line coincident with the conjugate axis.

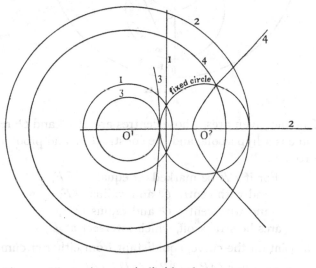

Fig. 40. Illustrating certain limiting forms of the axial plane.

(2) When the two circles touch internally, it is a straight line, coincident with the transverse axis.

(3) When they touch externally there are no special characteristics in the resulting hyperbola.

(4) When the radius of curvature of the variable circle is equal to the distance between the foci plus half the radius of curvature of the fixed circle, the hyperbola is rectangular, i.e. its asymptotes are at right angles.

Prop. 19. **To find the axial plane between two arcs, both being in the same sense, either anticlinal or synclinal, but where the radius of one arc or both of them is greater than the straight line joining their centres.**

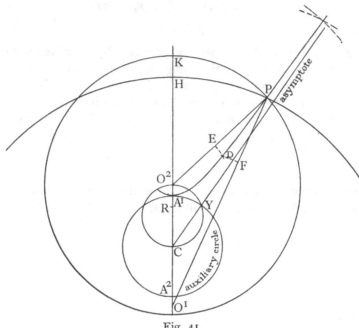

Fig. 41.

Let PH and PK be the given arcs, whose centres are at O^1 and O^2 respectively.

Then the axial plane is a hyperbola, and the construction and proof are very similar to that given in Prop. 16.

For if PE be marked off equal to PF,
and with centre O^1 and radius O^1F,
and with centre O^2 and radius O^2E,
arcs be struck off, which intersect at D,

then D is a point on the curve, equidistant from either circumference.

For $O^1P - O^2P = O^1D - O^2D$.

Proceeding to find the asymptotes of the hyperbola, which will be useful to us if the figure is reversed, and the arcs are considered to be both synclinal, we must bisect O^2R at A^1,

where $O^1R = PO^1 - PO^2$,
and $O^1A^1 - O^2A^1 = O^1R$.

Making O^1A^2 equal to O^2A^1, we construct the auxiliary circle on A^1A^2 as diameter,
and the second circle on O^2C as diameter.

Then CY is an asymptote.

As before, we may show that $2O^2A^1 = KH$.

$$\text{For } O^1A^1 - O^2A^1 = O^1R,$$
$$\text{and } O^1H = O^1P;$$
$$\text{also } O^2K = O^2P.$$
$$\therefore O^1H - O^2K = O^1R.$$
$$\text{But } O^1R = O^1O^2 - O^2R.$$
$$\therefore O^2R = O^1O^2 - O^1R.$$
$$\therefore O^2R = O^1O^2 - O^1H + O^2K$$
$$= O^1O^2 - O^1H + O^2H + KH.$$
$$\text{But } O^1O^2 = O^1H - O^2H.$$
$$\therefore O^2R = KH,$$
$$\text{and } 2O^2A^1 = KH.$$

The asymptote will hardly be required when we are dealing with the actual figure, showing two anticlinal arcs, both of which will ultimately cut out.

Proceeding now to consider another limiting case, we may imagine the distance between the centres of the arcs expanding. Let the centre of one of the arcs be fixed, while the other recedes to infinity. Then one of the arcs remains a circle, while the other becomes a straight line. We shall now show that the axial plane between them is a parabola, and this curve is consequently the limiting form of a hyperbola.

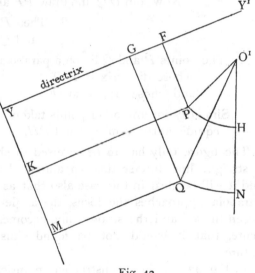

Fig. 42.

The following proposition may be applied where one flank of a fold consists of a great thickness of beds lying at the same dip, and constituting what is sometimes termed a monoplane (Figs. 42 and 43).

Prop. 20. To find the axial plane between two intersecting horizons, the one being an arc in the synclinal sense, and the other a straight line.

We may consider this proposition as one in which we are dealing with two synclinal arcs which intersect, the one with a finite radius, and the other with a centre that lies at infinity.

For let O^1 be the centre of the synclinal arc of finite radius, then O^2 lies at infinity, i.e. any straight line drawn perpendicularly to the given straight line PK is a radius, which meets all similarly drawn parallel straight lines at O^2 at infinity.

$$\text{Then } \infty - O^1P = \text{constant} = \infty.$$

$\therefore$ The axial plane required is a hyperbola one of whose foci is at infinity.

Now draw any perpendicular KM, and any radius O^1H to the arc PH and produce.

Mark off any distance KM on the perpendicular, and on the radius O^1H produced mark off HN equal to it.

With centre O^1 and radius O^1N draw the arc NQ, and draw MQ at right angles to KM, meeting NQ at Q.

Then Q, being equidistant from the straight line KP and from the arc PH, is a point on the curve.

Also along MK produced draw KY equal to O^1P, and draw YY^1 perpendicular to MY.

Then YY^1 is the directrix of a parabola, which has O^1 as focus, and which passes through P and Q, and is the axial plane required.

For a parabola is defined as the locus of a point which moves in the plane containing a given point and a given straight line, in such a manner that its distance from the given point is equal to its perpendicular distance from the given straight line. The given point is called the focus, and the given straight line the directrix of the parabola.

Now join O^1Q and draw PF and QG perpendicularly to YY^1.

Then $PF = O^1P$,
and $QG = O^1Q$.

$\therefore$ The points P and Q lie on a parabola
whose directrix is YY^1,
and whose focus is at O^1.

Similarly for any other points taken equidistantly from KP and PH.

The figure only has to be reversed to show a straight line intersecting an anticlinal arc, and it will be clear in this case also that as the parabola approaches the focus, the angle between the arc and the straight line becomes so acute, that it would not be found thus in nature.

In Fig. 43 we show constructed an anticline whose crestal dips give the circular arc HK; to the left of H the fold becomes a monoplane, and to the right of K the flank is synclinal, and is controlled by the centre O^1. Below P, the centre of the crestal arc, the axial plane becomes parabolic. The directrix YY^1 may be drawn at a distance from P equal to O^1P, and parallel to the straight line PQ.

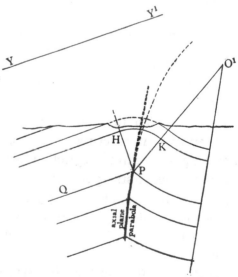

Fig. 43. Illustrating an anticline, one of whose flanks is a "monoplane." Below P the axial plane is a parabola whose focus is at O^1, and whose directrix is YY^1, above P it is a straight line bisecting the arc HPK. To show its form the parabola is produced above P.

We have now analysed the behaviour of the axial plane between two intersecting arcs either of like or of opposite form, and we may now pass on to consider the relationship between the various curves, which build up the axial plane of any competent fold in

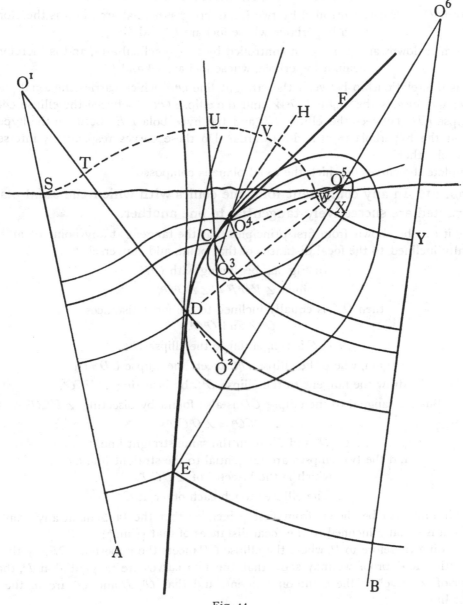

Fig. 44.

cross section, composed of a series of tangential and intersecting arcs. Reverting to Fig. 32, the form of which we here repeat in Fig. 44, we see that:

From O^4 to C the axial plane is controlled by an anticlinal arc intersecting a synclinal arc, and is therefore an ellipse, whose foci are O^3 and O^5.

From C to D it is again controlled by an anticlinal arc intersecting a synclinal arc, and is again an ellipse, whose foci are O^2 and O^5.

From D to E it is controlled by two intersecting synclinal arcs, and is therefore a hyperbola, whose foci are O^1 and O^5.

From E downwards it is again controlled by two synclinal arcs, and is therefore again a hyperbola, whose foci are O^1 and O^6.

What is the relationship between the straight line O^4F, which carries the axial plane to surface, and bisects the angle VO^4W, and the ellipse O^4C; between the ellipse O^4C and the ellipse CD; between the ellipse CD and the hyperbola DE; between the hyperbola DE and the hyperbola from E downwards? Do these curves respectively intersect or touch each other?

Complete the curves of which the axial plane is composed.

Prop. 21. In any competent fold, the conics with which the axial plane is constructed are successively tangential to one another.

Now it can be shown from first principles that the tangent at any point of an ellipse is equally inclined to the focal distances of that point, and conversely[*].

In Fig. 44, beginning with O^4F,

since $\angle VO^4F = \angle FO^4O^5$,

then O^4F is equally inclined to the focal distances

O^3O^4 and O^5O^4.

$\therefore$ O^4F is tangential to the ellipse O^4C.

Again, where the ellipse O^4C meets the ellipse CD at C,

draw the tangent to the ellipse O^4C by bisecting $\angle UCO^5$.

But the tangent to the ellipse CD is also found by bisecting $\angle UCO^5$.

$\therefore$ $\angle O^3CO^5 = \angle O^2CO^5$.

$\therefore$ O^2, O^3 and C are in the same straight line,

and the two ellipses are tangential to the straight line CH,

which is the bisector of $\angle UCO^5$.

$\therefore$ The ellipses touch each other at C.

Now it may also be shown from first principles that the tangent at any point on a hyperbola is equally inclined to the focal distances of that point[†]:

Proceeding therefore to D, where the ellipse CD meets the hyperbola DE, by the same construction and proof we may show that the two curves are tangential at D, that the bisector of $\angle O^1DO^5$ is the common tangent, and that O^1, D and O^2 are in the same straight line.

Similarly the hyperbola DE touches the hyperbola below E at E. Therefore the curves of which the axial plane is composed are successively tangential to one another.

* It is intended only to enunciate those propositions which can be found proved in any text-book. As here, C. Smith, *Geometrical Conics*, 1926, p. 85.   † *Op. cit.* p. 126.

Had one of the curves of the axial plane been controlled by a straight line instead of by an arc, thereby making it a parabola, it can be seen that this conic would have been

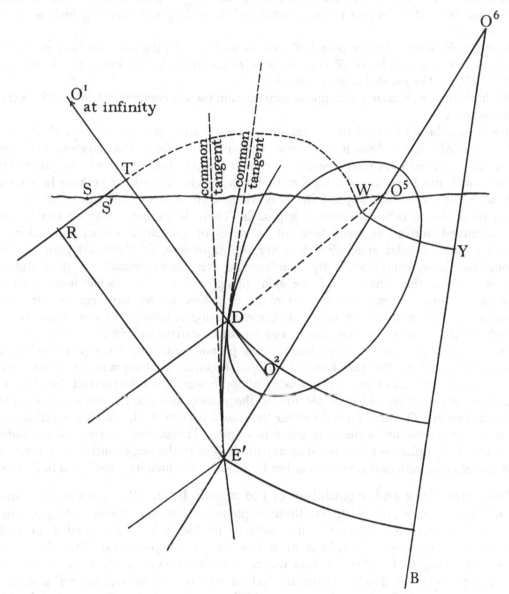

Fig. 45. This is Fig. 44 repeated, except that O^1 has receded to infinity, making one flank of the anticline a "monoplane." Note that the ultimate parabola of which the axial plane is formed bends through the vertical and back towards what is the "steep limb" at surface. The effect of the steep dips around W is entirely counteracted by the great thickness of beds exposed in the "monoplane" on the opposite side.

tangential to those adjacent to it also. For let O^1 recede to infinity along the line TO^1, so that the arc ST becomes the straight line, $S'T$ (Fig. 45), then DE' and the curve below E' become parabolas with O^5 and O^6 as respective foci. The bisector of the angle O^1DO^6 still remains the common tangent at D, for it is equally inclined to the focal distances DO^1 and O^5D of the parabola DE', as also to the focal distances O^2D, O^5D of the ellipse DC. The fact that O^1 has receded to infinity does not affect the inclination of the radius O^1D.

Again at E', where the parabola DE' cuts the radius O^6E', the common tangent to the parabola DE' and that below E' is the bisector of the angle $RE'O^5$, where RE' is parallel to DT. Hence the parabolas are tangential*.

We have thus worked out a complete construction for any competent fold, and its axial plane in section.

It will have been observed that our analysis of the matter has embraced practically the entire field of what is known as "conic sections," the ellipse, the parabola and the hyperbola and their inter-relationships. A clearer grasp of the subject will be gained by any who will turn to such a text-book as that which we have cited, for we have here only dealt with the curves and applied the laws which control them as occasion has arisen. There are of course other methods of approach. There is no reason why we should not treat a curved horizon as being bent into a series of tangential conics, for instance, instead of into circular arcs, when, however, the expression for the axial plane would become needlessly complicated. By using circular arcs, we can visualise in the simplest manner possible the behaviour of the axial plane, and we can decide from a mere inspection of our evidence collected from the field how far we have reached finality. When our evidence shows an anticlinal curve meeting a synclinal curve, there is no finality, for the axial plane is elliptical, and cannot be carried indefinitely to depth: but let the anticlinal part of the curve become less pronounced, until it is represented by a straight line, until, in fact, this flank of the fold becomes a monoplane; then the axial plane becomes a parabola, expressing finality at depth, were it not for the fact that there is a limit in nature to the width of outcrop of the monoplane and the thickness of beds involved therein. The final form for either limb *must* be synclinal, when the axial plane becomes a hyperbola and extends to depth to infinity. It becomes a curve, which if both synclines are complete is not affected by any other folds in the neighbourhood. Evidence must therefore be collected far out on either limb before the final hyperbola can be drawn.

The axial plane and attenuation of the middle limb. Where the middle limb of a fold is attenuated, the position of the axial plane becomes modified accordingly. The fold then becomes "incompetent," since horizons no longer maintain parallelism, and though there may be ample evidence at surface for the determination of the degree of attenuation along that surface, it does not necessarily follow that the same degree of attenuation occurs at depth. There are indeed so many unknown factors, and the complete treatment of the problem geometrically becomes so complicated, that in the present state of our knowledge, such treatment is rarely worth while.

---

* The second focus of a parabola being at infinity, this proposition may be enunciated as follows: The tangent at any point of a parabola bisects the angle between its focal distance and the perpendicular therefrom to the directrix. This is a corollary from Prop. 20, Fig. 42. See also C. Smith, *op. cit.* p. 39.

In the first place, our definition of the axial plane for a competent fold no longer holds. It is no longer the locus of a point which moves equidistantly from either limb of the fold as defined by any particular horizon, nor is there, apparently, any fixed ratio between the axial distances from either limb as defined by each horizon in turn. The best we can do is to join up the apical points after constructing the fold for various horizons, and call the curve produced the section of the axial plane. A fold with an attenuated middle limb does, however, render us some evidence at surface, which will give us a fair idea of its construction at depth. In the first place, the ideal similar folding illustrated in Fig. 9 is never realised, as all folds die out eventually at depth, and there is nearly always variation in the attenuation of the middle limb, some rock zones becoming more affected than others. There is, also, generally a tendency for the attenuation regularly to reach a maximum in the central area of the middle limb, and for a regular dying away of the attenuation towards the axes of the anticline and its foresyncline respectively.

Fig. 46 shows a somewhat idealised though typical fold with a vertical middle limb, whose maximum attenuation is by 50 per cent. dying away towards each axis. Horizons

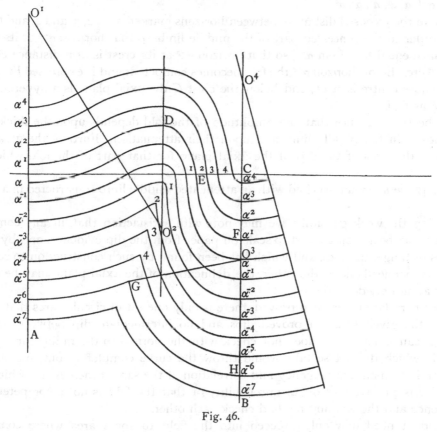

Fig. 46.

are shown for unit thickness. The gentle limb is competent, as is also the trough of the foresyncline, and this part of the fold is controlled by arcs whose centres are at O^1, O

and O^4. What surface evidence have we for the drawing of the fold at depth? Radii O^1A, O^1O^2 and O^4B are definitely fixed by known dips. Moreover, we know that there will be no attenuation along the radius O^4B. We know also the correlation across the fold of horizons α^{-1}, α, α^1, α^2, α^3, α^4. From C therefore downwards mark off unit thicknesses from α^4 along the radius CB, and repeat the process for radius O^1A. Horizons α^{-2}, α^{-1}, α, α^1, α^2, α^3 and α^4 may be drawn from the gentle flank to the crest, and horizons lower than α^{-2} may be drawn for part of the way. We must now join the horizons with the known point of intersection with surface in the middle limb with the most reasonable curve that we can find. Thus, for horizon α^2, the normal and dip of this horizon is given at D, and again at E and F. We may draw the curve through these points freehand, bearing in mind that it must be at right angles to the normals which pass through them. We could draw this curve geometrically by using a construction similar to that for the interpolated dip, but no two horizons have the same curvature, and the process repeated for each horizon would be unnecessarily laborious.

Complete the construction of the exposed portion of the middle limb freehand for horizons α^{-1}, α, α^1, α^2, α^3, α^4.

Now take the exposed distances between horizons marked as 1, 2, 3, and 4, and transfer them in order to the concealed part of the middle limb, so that horizon α^{-2} at its crest is at a distance equal to 1 from α^{-1}, so that horizon α^{-3} at its crest is at a distance equal to 2 from α^{-2} etc. Below horizon α^{-5} the fold becomes competent and is expressed by a single arc GH whose centre is at O^3, and below the crest G the axial plane is a hyperbola with O^1 and O^3 as foci.

It will be seen at once that the amplitude of the fold depends upon the thickness of beds exposed in the middle limb and upon their attenuation. Further, the greater the attenuation the greater the dip of the axial plane for that part of the fold which is so affected.

In Fig. 47 we show an overfold with an attenuated middle limb constructed in a similar manner.

Reviewing the work generally, we may note with satisfaction that, in any competent fold, which has been constructed to scale on paper from the dip evidence, and by means of a series of tangential arcs, and which has been found by such construction to correlate correctly from one flank to the other, the delineation of the axial plane may be mathematically accurately defined.

For once correlation has been proved, there is only one set of circular arcs that can be drawn for the given series of proved dips, and no intermediate dip between any two given dips can exist, which does not check with the normal to the radius through that dip. A dip which did not so tally would throw the curve completely out, and it would be impossible to maintain tangency and correlation at the same time. A dip which does not so tally is proof either of its unreliability or that the fold is not a competent one. The evidence and the working method check each other.

The student need now only proceed into the field to some area where competent folds are known to exist, and he will be delighted with the simplicity of these methods of construction, and with the accuracy with which the evidence checks in. Unfortunately, competent folds of reasonable dimensions are rare in England, and where they do exist the essential evidence is obscured by the works of man or by vegetation. In Chapter VI therefore we give examples from other countries, and show how the geometric method may be applied. Where all the evidence has been mapped there need be little argument

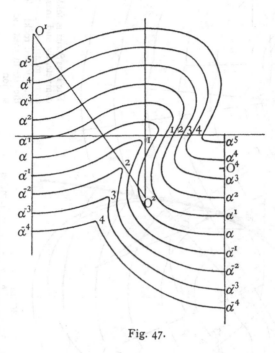

Fig. 47.

about the position of the axial plane in section, and none at all where the folding is competent, as is often the case, when it can be drawn with perfect accuracy as a smooth curve, composed of a number of tangential conics.

We append (Fig. 48) an example of competent folding taken from South-western Persia in the Bakhtiari mountains. Owing to the complexity of the figure, for only two of the anticlines in the section has the axial plane been constructed, but the curves of which they are composed are shown in full. The student can trace for each axial plane the position in which each conic touches the next at the point of intersection of the relative radii, and the centres of each tangential arc are numbered consecutively. The anticline marked "Jeriveh," which pitches up to the north-west, is shown in the frontispiece sketch. It will be noted from the scale that some of the radii do not meet lower than 12 miles above surface, thus emphasising the necessity for small scale work in big mountain ranges, in order that the construction lines may be brought within reasonable limits. From the nature of the folding of these rocks, there is no reason to suppose that competency should not continue almost indefinitely downwards, and it will be noted that at a depth of some ten miles below surface the folds are scarcely represented at all, and that there is merely a general dip from the continental platform

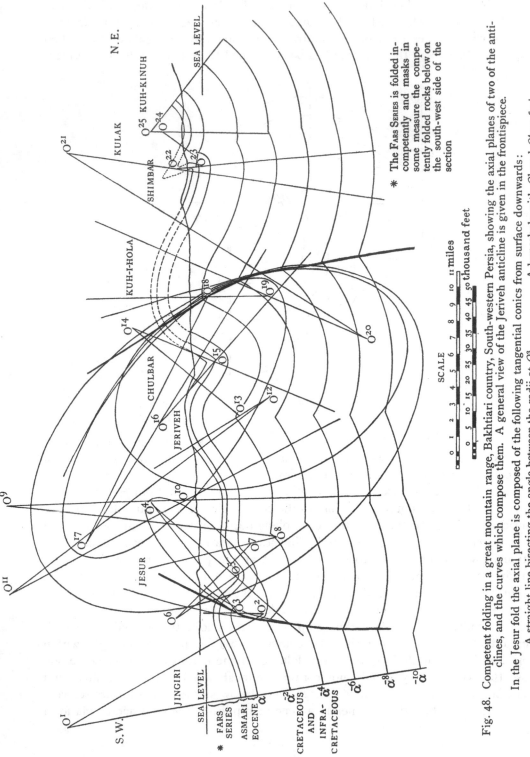

Fig. 48. Competent folding in a great mountain range, Bakhtiari country, South-western Persia, showing the axial planes of two of the anticlines, and the curves which compose them. A general view of the Jeriveh anticline is given in the frontispiece.

In the Jesur fold the axial plane is composed of the following tangential conics from surface downwards:

A straight line bisecting the angle between the radii at O^3.
A hyperbola with O^1 and O^4 as foci.
An ellipse with O^2 and O^4 as foci.
A hyperbola with O^1 and O^5 as foci.

In the Kuh-i-Hola fold the axial plane is composed of the following tangential conics :

A straight line bisecting the angle between the radii at O^{18}.
An ellipse with O^{16} and O^{20} as foci.
An ellipse with O^{17} and O^{19} as foci.
A hyperbola with O^{16} and O^{21} as foci.
An ellipse with O^{17} and O^{20} as foci.
A hyperbola with O^{15} and O^{21} as foci.

* The Fars Series is folded incompetently and masks in some measure the competently folded rocks below on the south-west side of the section

on the north-east. Speculation as to what actually does happen in the infra-Cretaceous rocks at a depth of a few miles would lead us to theoretical arguments which are no part of this work, but it may certainly be said that the solution shown is by no means an impossible one*.

* Cf. in this connection, Leith, *op. cit.* p. 125, where the work of Chamberlain and Salisbury and others is reviewed. Whether or not his theory is acceptable, Wegener has collected much important data bearing upon the depth of earth folding. Translation by J. G. A. Skerl, *The Origin of Continents and Oceans*.

CHAPTER V

GEOLOGICAL MAPPING FOR ACCURATE
SECTION DRAWING

IN an unsurveyed country the preparation of a geological map, which may be used afterwards for the purpose of section drawing by geometrical methods entails very careful work. Any error in topographic survey, or errors brought about by accidental transgression of geological horizons, will be immediately discernible by apparent variations in thickness in the line of the sections taken. In fact, map and sections check themselves. The geologist must be on his guard against assuming that apparent thickness variations that occur on his map are actually present, and he must always be ready to revise his map in the field, in order to eliminate any possible sources of error.

The question of thickness between the beds he is mapping should always be before the mind of the geologist as he traces his horizons. If variations are found, they are certain to follow a definite rule, and to be of definite intensity. The angle made in section between two horizons, which are approaching one another by overlap, has its limit in magnitude, and in marine or lacustrine deposits can never exceed a degree or two. Thus, taking South-west Persia as an example of a country in which there occurs a highly variable series of river deposits of Pliocene age, laid down during contemporaneous folding, the angle between two adjacent mapped horizons, which approach and ultimately meet, has been found to reach an extreme of no more than 3°. The exaggerated conception of the angle that is possible in nature has its foundation probably in the many diagrammatic and out-of-scale figures that appear in text-books of geology.

The question of scale should be considered in relation to the amount of detail it is proposed to show, and the ease in handling the sections that it is proposed to draw. Where changes of dip are slight, curves of large radius will come into the section, and mapping may more conveniently be carried out on a smaller scale than when changes of dip are large. It is seldom that one need use a scale greater than that of six inches to a mile, and a scale of three inches to a mile is generally large enough, even for the fullest detail. In exploratory work a scale of one inch to a mile need never be exceeded.

The plane table can be used with advantage everywhere, even in dense forest about which we shall have more to say later. The plane tabler* should know exactly what are his limits in the accuracy of his drawing or of his survey. A practised surveyor should be able to work to about a five-hundredth of an inch on paper. The same amount of detail per square inch of paper will be inserted, whatever the scale is. Thus in well-established plane table surveys, like those of India or the United States, the actual

---

* The student is supposed to be fully acquainted with the use of the plane table and with the theory and practice set out in (say) Close and Cox, *Text Book of Topographical Surveying*, Chap. VIII (H.M. Stationery Office), or Wilson, *Topographical Surveying*, Chaps. VII, VIII, IX (Chapman and Hall).

scale is only discernible by reference to the scale on the map, the same amount of detail per square inch of paper appearing on all the various scales used. It will thus be seen that a map on a scale of 6 inches to a mile will take thirty-six times as long as one on a scale of one inch to a mile. In fact, on all scales between 6 inches to a mile and half an inch to a mile, where the factor of rate of travel is barely noticeable, the time taken to survey a given area varies directly as the square of the scale. This, of course, is an important point to bear in mind, and scale should be carefully chosen in relation to the time at the disposal of the surveyor.

Although it may help the geologist considerably to have at his hand a trigonometric triangulation upon which to base his work, a special survey of this nature is not really necessary and is very expensive, as a separate expedition would have to be undertaken for this work. A great deal of time is wasted by the rule of thumb surveyor on refinements which are in reality quite irrelevant to the geological work in hand. In a survey, for instance, embracing an area of a thousand square miles, a plane table triangulation, which ignores the curvature of the earth from station to station, or which, by discrepancies due (say) to the unequal shrinking of joined sheets, gradually introduces an error, which may amount to distortion at the edges of the map to as much as half a mile, is just as useful to the geologist as a map produced by geodetic survey. It is not error over large distances by distortion that will affect our section drawing, but inaccuracy of detail immediately around the neighbourhood with which we are at the moment dealing.

The geologist should train himself to distinguish between those parts of the map where refinement is required and those parts where rough sketching and form-line work can be indulged in without detriment to the work as a whole.

Plane Table Triangulation. Having settled the question of scale, which in nearly all preliminary work need not exceed that of one inch to a mile, the geologist will begin his plane table triangulation, bearing in mind that accuracy for each plane table sheet is essential, but that a moderate and inevitable degree of distortion over large distances is immaterial. The preliminary part of the triangulation is the only stage that need be worked apart from the geology, and this is repeated sheet by sheet, a new base line being measured for each sheet. The dimensions of the plane table should be about 2 feet by 1 foot 6 inches, and it should be fitted to rigid, non-collapsible legs, and with a Johnson cup and ball-fitted head.

It is unnecessary to triangulate the whole of each sheet, and the following method gives very accurate results:

Measure a base line, which on a scale of one inch to a mile should be about 5000 feet in length; extend this by setting up the plane table at either end, orienting and drawing rays to two points abeam of the base line such that they subtend an angle from either end of not less than 45°. (Fig. 49.) From both ends of the base line, rays are drawn to all prominent points expected to appear on the sheet, and extensions of these are drawn in at the edge of the plane table to facilitate orientation. The magnetic north point is carefully drawn in across the paper as soon as the plane table is in orientation at the first station.

Some practice and experience are necessary in "shooting" the various prominent points chosen, as there is rarely time to travel round erecting cairns upon them. Choose, if possible, a prominent rock at or very near to the summit, and which there is good reason

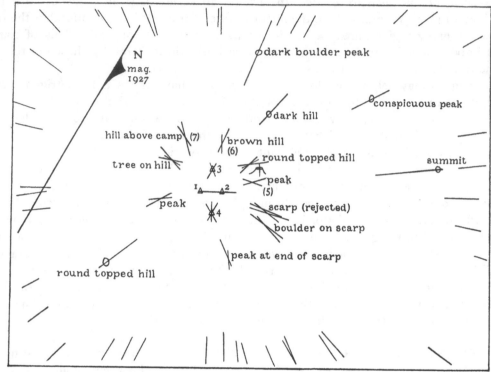

Fig. 49. The appearance of the plane table after the first three stations have been occupied. Only stations 1, 2, 3 and 4 should have been inked in. From station 1 to station 2 is the measured base line, and from station 3 to station 4 is the base line extended by triangulation. The stations would be then visited in the following order, if satisfactory intersections were obtained, namely 4, 5, 6 and 7, after which intersections outside the inner ring shown will have been obtained. The rest of the sheet is surveyed by resection. The rays round the edge of the sheet are for purposes of orientation. The scale may be anything.

to believe is visible from all sides; but where the top of a hill is rounded or flat, arrange that the vertical wire of the instrument intersects the apparent centre of the hill in each position, as in Figs. 50 and 51, and, provided that the top of the hill is not too irregular, this will give an intersection very near the top of its rounded or flat summit.

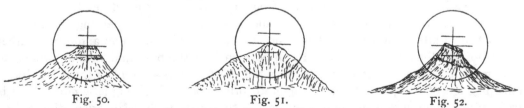

Fig. 50.   Fig. 51.   Fig. 52.

Figs. 50, 51, 52. Showing the right way to sight points of various types.
The image in the instrument is of course inverted.

A small dipping outlier is better intersected through its centre than through its highest point (Fig. 52). Beware of an oblique view of an escarpment; the highest point does not always appear to be the highest from a side view, and may not be always visible (Fig. 53).

Fig. 53. Showing two views of the same scarp, which would give poor intersection. The position of either *A* or *B* in either figure is poorly marked.

The observer may now proceed to triangulate the centre of his sheet, but it is unnecessary to cover more than half the area of the paper, for the outer points may be found by resection. At the edge of the sheet three points are determined, which are common also to the sheet adjoining, or, if other sheets are to be connected, three common points will be required to make each connection. As soon as the triangulation of the central portion of the sheet is completed, geological mapping may go hand in hand with the resection of the rest of the sheet. Outside his central triangulation points proper, the observer should not have to visit any one resected station more than once.

As soon as the first sheet is completed the geologist moves camp to the middle of the next sheet, measures a fresh base line, triangulates the central part of the sheet, and intersects the points common to the adjoining sheet or to each adjoining sheet. The

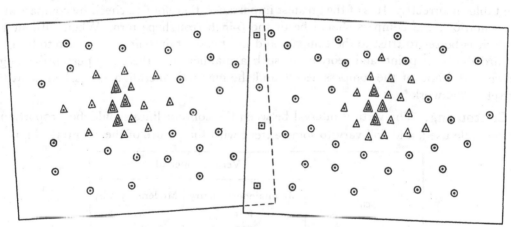

Fig. 54. The building up and joining of two plane table sheets.
△︀ Base line and extended base line stations.
△ Intersected stations.
⊙ Resected stations.
□ Resected connecting stations.

geology is again surveyed as soon as the triangulation proper of the central part of the second sheet is completed.

The two sheets are thus built up as is shown in Fig. 54.

Resection and the three point problem. The triangle of error found on resection when the plane is not exactly in orientation is best dealt with in Wilson's *Topographic Surveying*, p. 186, and the figure given there should be carefully examined. For a detailed consideration of the three point problem the student is referred to this work. It is sufficient here to set out the four simple rules for the reorientation of the table for the elimination of the triangle of error; they are:

(1) When the point required is on or near the circle passing through the three given points, its position by resection is indeterminate, and another point, off the circle, from which to resect, must be chosen.

(2) When the point required is within the triangle formed by the three given points, its position is within the triangle of error.

(3) When the point required is without the triangle but within the circle formed by the three given points, orient on the middle point, then its position is on the side of the line from the middle point remote from the point of intersection of the other two lines.

(4) When the point required is without the circle through the three given points, orient on the most distant point, then its position is always on the same side of the line from the most distant point as the point of intersection of the other two lines.

The observer, resecting outwards from a plane table triangulation, will find that this last rule is the most frequently called into requisition. It must, however, be handled with the greatest care. For if any of the triangulation points used for resection are slightly out of position, not only will there be a triangle of error when the plane table is correctly oriented, but it will often be possible to obtain a single point intersection by orienting the table incorrectly. It is of the greatest importance, therefore, to check by compass after reorientation; the compass should be of the 6-inch trough pattern. When without the circle it is better to trust to the compass and use two certain points only than to bring in a third doubtful point, and reorient in such a manner that there is a perceptible angle between the line of the compass needle and the magnetic meridian as marked in, at the outset of the work*.

Contouring. The vertical interval between the contour lines should be proportional to the scale used, and will vary to some extent with the nature of the country. Thus:

	Vertical Interval	
	---	---
Scale	Mountainous country	Moderately hilly
	feet	feet
$\frac{1}{2}$ inch to a mile	500	200
1 inch to a mile	200	100
3 inches to a mile	100	50
6 inches to a mile	50	20

It is better to use multiples of ten than to introduce the 25 feet or 250 feet vertical interval, since if more detail is required later on, the odd tens can be interpolated.

* In this connection beware, of course, of the presence of any possibly magnetic rocks such as basic dykes. The effect of such rocks is generally quite local.

The contouring and mapping of the topography generally can be carried out simultaneously with the geological mapping. For both, the Beaman Stadia Arc fitted to the telescopic alidade is of the utmost value. The observer resects, if possible, upon a horizon that is being traced, and determines the altitude of his station either by multiplying the distance, as given by the map, with the sine of the angle from a known station, or by multiplying direct with the factor shown on the Beaman arc. He sends down his rodman to any geological horizon that he is tracing in the valley, sights on to the rod to get his distance, and multiplies this by the factor on the Beaman arc, which automatically compensates for the fact that the rod is held vertically, and not square on to the diaphragm of the instrument.

The ordinary rod is not suitable for this operation, as it is difficult to read at distances as great as 2000 feet. A stout 14-foot hinged rod, divided into feet and quarters of a foot, and with each alternate foot painted red and black respectively, gives the best results, and it is unnecessary to paint on the rod the numbers of feet, as these can be readily counted. With such a rod, for distances over 1400 feet, the reading is taken between one of the stadia lines and the horizontal cross wire, and then doubled. One or, at the most, two readings on the rod is all that is generally required from each station.

Mapping geological evidence. Dip readings are taken as frequently as possible, and entered on to the map. Where there is rolling dip or uncertainty in its exact value, the general dip should be entered, and, if necessary, differentiated from those readings which are absolutely reliable. The number of dip readings entered should be at least as many as 5 per square inch of paper. They should be read by sighting transversely along the edge of a clinometer fitted with a plumb-bob and a fixed dial. After some practice the observer will find that he is able to estimate by eye dip readings to the nearest degree, and that only an occasional use of the instrument will be necessary.

The thickness of strata between mapped horizons should be constantly checked, and should variations occur the observer must discover why this is so. By going over the ground again, he may find:

(1) That he has transgressed horizon owing to the poorness of the exposures.

(2) That he has transgressed horizon by passing across a dip or tear fault unnoticed.

(3) That there is structural attenuation due to increase of dip or overfolding.

(4) That there is actual stratigraphic variation in thickness.

In no case must a variation in thickness be left unexplained, nor must a fault be passed over without a determination of its throw and displacement, and whether it diminishes or increases in throw in either direction. If stratigraphic variation is present it always follows some rule, the elucidation of which will be of prime importance to the correct drawing of sections across the map.

In order to apply rapidly a check to thicknesses in the field, the author has devised a scale (Fig. 56) by means of which the width of outcrop in a horizontal plane may be read straight off for various observed angles of dip. This scale may be drawn off in colours, each colour representing the angle of dip to which the vertical stroke refers.

hinge

Fig. 55. Rod suitable for rapid plane table work with the Beaman Stadia Arc. It is divided into feet and quarters of a foot, and so reads to twenty-five feet between the stadia wires.

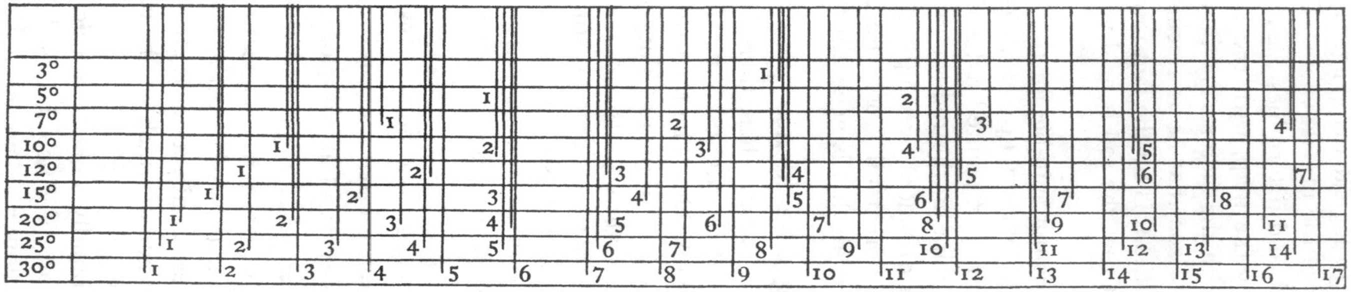

Fig. 56. Scale of one inch to a mile to show width of outcrop in a horizontal plane for various angles of dip up to thirty degrees and for various thicknesses of strata. The divisions are shown for every thousand feet of thickness. In practice the vertical strokes are drawn in different colours for the angle of dip to which they refer, and the scale is cut out of mounted drawing paper.

Should a mapped horizon disappear under alluvium, or for any other reason be badly exposed, it is permissible to *interpolate* it to accord with the topography and the surrounding evidence generally, from those horizons immediately above or below it. Where this is done, the horizon is shown as a dotted line only, and the word "interpolated" inscribed along it. Except for special purposes, or perhaps along large stretches such as a coastal plain, where all the evidence is obscured, it is unnecessary to map alluvium, and by not doing so a great deal of time is saved. The observer should remember that it is "solid" work that is wanted, and that if he, working in the field, cannot interpret what is obscured by alluvium, no one else can do so. *All* the evidence that he finds must be included, and, when there is doubt, or more than one possible interpretation, a note should be made to that effect upon the map. A map should, in fact, be so complete that a report upon it becomes redundant. It should tell the whole geological story.

In Fig. 49 a plane table sheet in its early stages with construction lines and notes is shown. The strokes round the edge of the sheet are placed there, when the alidade is in position, for the purpose of reorientation. Soon after the stage that is illustrated in the figure has been reached, the observer may proceed to resection and geological mapping.

Figs. 84 and 88 are fair examples of a finished plane table sheet slightly reduced, though in practice streams are coloured blue and contours brown. The geology also is coloured in and not shaded.

Plane Table Survey in Forested Country. The plane table can be used with advantage in forested country, provided that the climate during the field season is a dry one.

Triangulation is no longer possible, since stations will be rarely open to view. The observer will have to resort to closed traverses. If a trigonometric survey is already in existence, traverses may, of course, be closed on triangulation stations.

Set up the plane table on the point from which the traverse is about to start. Sight to a rodman, placed in any convenient position on the line of the traverse; decide on the orientation of the table, draw the ray, mark in the magnetic meridian on the table, and plot the position of the rodman. Proceed to a convenient point beyond the rodman, orient the table by compass, and check with a back sight on the rodman. Measure off the position of the rodman, and plot the position of the table; sight on to a second rodman placed beyond the table.

Distances are, of course, more accurate when taped.

Elevations may be calculated from the Beaman arc, or the angle on the vertical circle of the alidade.

A complete map is usually impossible to construct within reasonable time by traverse methods and traverses will generally follow along the more important geological lines, and along those stream channels where the evidence is best exposed. Fig. 57 is an example of a map constructed from such a series of traverses, but altitudes are not shown.

Section Drawing. After a map has been completed by plane table triangulation in open country, it should not be necessary to revisit the ground in order to draw a section. Such a map should bear the test of sections drawn across it anywhere. In forested country, where contour work is impossible, however, it is generally necessary to construct the profile, and measure up the section direct in the field.

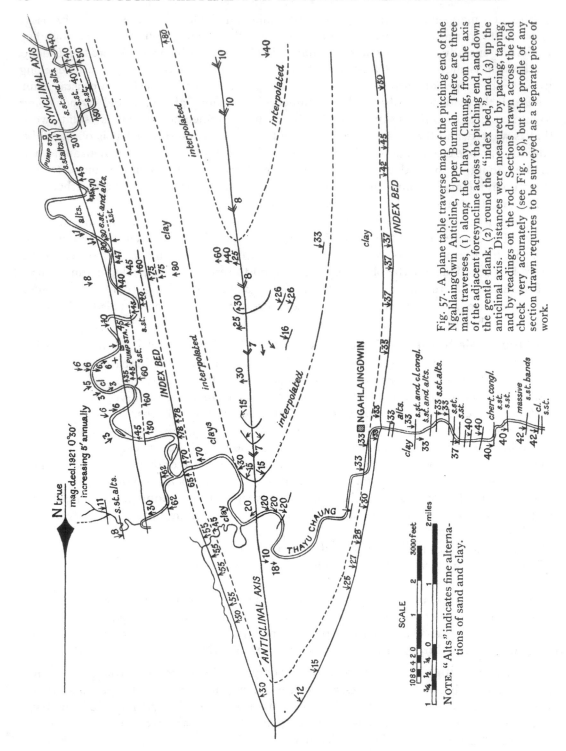

Fig. 57. A plane table traverse map of the pitching end of the Ngahlaingdwin Anticline, Upper Burmah. There are three main traverses, (1) along the Thayu Chaung, from the axis of the adjacent foresyncline across the pitching end, and down the gentle flank, (2) round the "index bed," and (3) up the anticlinal axis. Distances were measured by pacing, taping, and by readings on the rod. Sections drawn across the fold check very accurately (see Fig. 58), but the profile of any section drawn requires to be surveyed as a separate piece of work.

NOTE. "Alts" indicates fine alternations of sand and clay.

CHAPTER VI

FLEXURES OF TERTIARY AGE IN THE PETROLIFEROUS ROCKS OF SOME EXTRA-EUROPEAN COUNTRIES

EXCEPT in those countries, perhaps, which have organised geological surveys, the greatest activity in geological mapping at the present day takes place under the direction and auspices of various oil companies, and has for its prime object the finding of oil. It is for this reason that, the distribution of petroleum being amongst other things subject to hydrostatic law, the attention of professional workers is directed so largely to earth flexures, their form, and the interpretation of their form from surface evidence. The bulk of this work, probably more than nine-tenths of it, for reasons of competition, remains unpublished; hence, in the journals of the day, the subject has not really attained its proper place.

The young student in England, therefore, who intends to take up petroleum geology as a profession, is handicapped at the outset by lack of publications dealing with the subject, and by the impossibility of obtaining in this country practice in field work which can be said in any way to be truly representative of what he may expect to meet in the petroliferous areas of the world.

To begin with, he is almost certain to find that during the greater part of his professional life he will be mapping folding or faulting or both in Tertiary rocks. There is no folding of the Tertiary rocks in England of any great significance, and what there is is poorly exposed. He will probably be dealing with extremes in the way of surface obstructions by vegetation or total lack of it, as with extremes of climate; he will probably be either in a forested country, where each item of evidence is extracted only under the utmost difficulty, or in a desert, where he will meet with a wealth of exposures (sometimes everything that *can* be exposed, *is* exposed) beyond anything about which he has ever dreamed. In the first country that he visits, if it is a desert, he is certain to become dogmatic, and his work will include dangerous theoretical generalisations from the evidence that he discovers in that country. In the second country that he visits, if far removed from the first but equally well exposed, he will find such a contrast to the first that he will hastily conclude that they have nothing whatever in common. After ten years of work in many other countries, he will begin to realise that variation in tectonic form is infinite, and that it is far beyond the power of any single worker to cover and to comprehend more than a small fraction of the whole. Although he will no longer generalise at random, he will apply those principles which he has found to be reliable to each of the areas that he has visited in turn.

Of those countries which have come under the author's direct observation, and which offer a contrast in structure, we will consider the Tertiary flexuring of three in some detail.

(1) **Burmah,** a region of simple folding and fairly simple stratigraphy.

(2) **South-west Persia,** a region of more complex folding due partly to the varying resistance of the rocks to earth stresses.

(3) **The west coast of the Peninsula of Sinai,** a region of simple normal faulting.

In spite of structural contrasts, each of these three countries has certain characteristics in common, and to all of them it has been possible to apply either in a greater or less degree those methods of section drawing from the mapped evidence, which we have described earlier. In all three countries folding and faulting have taken place under no very considerable load, and in all of them there is a succession of rocks of varying resistances, though this is much emphasised in Persia. In all these countries there is a marked échelon arrangement of the structures as opposed to virgation.

We will take Burmah first as a country affording good examples of simple folding, to which may be readily applied the geometrical constructions detailed in Chapter IV.

BURMAH

In Burmah* there is a succession in which clays and rhythmically intercalated sandstones (Pegu Series) lie below a more rigid series of massive sandstones (Irrawadi Series), though the junction between the two is not a true time horizon. The sandstones traced from north to south vary in succession from below into clays and sandstones of the Pegu type. There is in fact here the record of a delta advancing from north to south throughout the time that the two series, which are Miocene and Pliocene in age, were being deposited. The variation is, however, as a whole regional, and in each individual fold it is generally possible to trace horizons across from limb to limb by mapping down the pitching end of the fold. One notable exception to this treatment is the Yenangyaung oil field itself, in which at surface the beds are so highly variable, and are moreover so poorly exposed, that accurate mapping is a great difficulty. This fold, however, is a very simple one, being very nearly symmetrical, with dips that rise to 40° on either flank. The axial plane is therefore vertical, the oil-bearing strata being distributed symmetrically on either side.

We may note in Burmah a general tendency, where the soft clays of the Pegu Series are exposed at surface at the crest maximum or core of the fold, to a considerable degree of pinching up into a very sharp flexure at the crest, where dips may rise to over 70° on the one flank, and may become inverted upon the other.

Fig. 58 shows the chief anticlinal axes of the oilfield region. The anticlines are comparatively sharp folds, separated by broad synclinal areas. The steep limbs of the anticlines are generally, though not always, on the eastern side of their respective axes. On the western side of the area a complex of metamorphic rocks termed "Axials" crops out from under the Tertiaries. A section through the area (Fig. 59) drawn on a similar scale shows the amplitude and extent of the folding, and illustrates how widely separated are

* For stratigraphy see L. Dudley Stamp, "The conditions governing the occurrence of oil in Burmah." *Journ. Inst. Pet. Tech.* vol. XIII, no. 60.

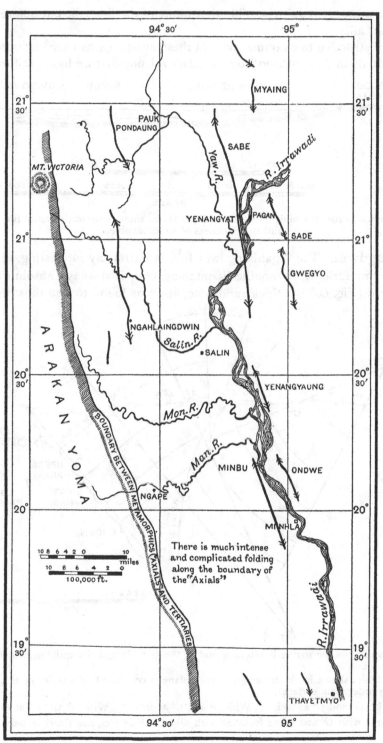

Fig. 58. Map showing the principal anticlinal axes in the oilfields region of Upper Burmah.

the anticlinal axes. It is quite possible that many of the folds die out before the "Axials" are reached at all.

It will be instructive to examine some of these anticlines, as they illustrate the validity of those methods of construction from surface evidence that we have set forth heretofore.

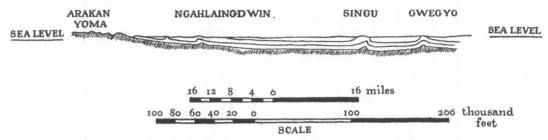

Fig. 59. Section across the oilfields region of Burmah, to show comparative amplitude of the folds, and the thickness of rocks involved.

Ngahlaingdwin. The Ngahlaingdwin fold is extremely interesting, in that, though distinctly asymmetric, at its southerly pitching end, it displays absolute competency. The index bed (Fig. 60) is a thick sandstone, and was found to be a most useful horizon,

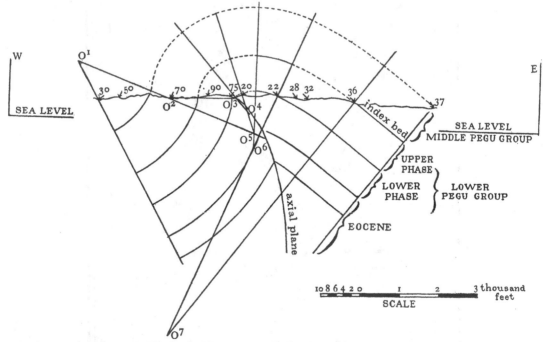

Fig. 60. Section across the southerly pitching end of the Ngahlaingdwin anticline, a competent fold in clays and sandstones.

Within the limits shown in the drawing the axial plane is composed of a series of tangential ellipses in the following order from surface:

With O^2 and O^4 as foci,   With O^2 and O^5 as foci,   With O^2 and O^6 as foci,

a very short curve with O^2 and O^7 as foci, and with O^1 and O^7 as foci, and finally a parabola with O^1 as focus.

and when the fold came to be drawn in section, on a very much larger scale than shown, by means of centres O^1, O^2, O^3 ... O^7, the correlation of this bed across the axis by geometric construction, and by mapping round the pitching end, tallied with mathematical exactitude. The drilling of certain wells near the crest bears testimony also to the accuracy with which the axial plane has been determined. It will be noted from the drawing that the final curve of the axial plane, as far as it goes in this figure, is a parabola with O^1 as the focus.

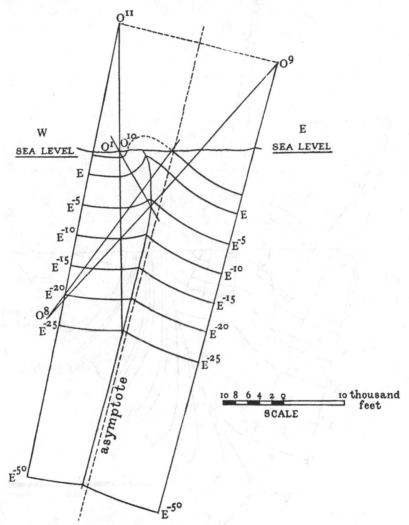

Fig. 61. The southerly pitching end of the Ngahlaingdwin anticline on a smaller scale, showing how the axial plane finally hades to the west in a hyperbolic curve, which has O^{11} and O^9 as foci.

The fold is even still more instructive if the axial plane is traced to greater depth, by taking evidence from further out on either flank. For it will be noted from Fig. 61 on a very much smaller scale, that it bends through the vertical, and finally hades in the opposite direction, to the west. This, it can be seen, is due to the fact that the general dip of the whole series is easterly. The fold (Fig. 61) is carried to depth by means of horizons drawn at 5000-foot intervals from the top of the Eocene (horizon E). Somewhere or other, of course, within the limits of the section, the metamorphics will be

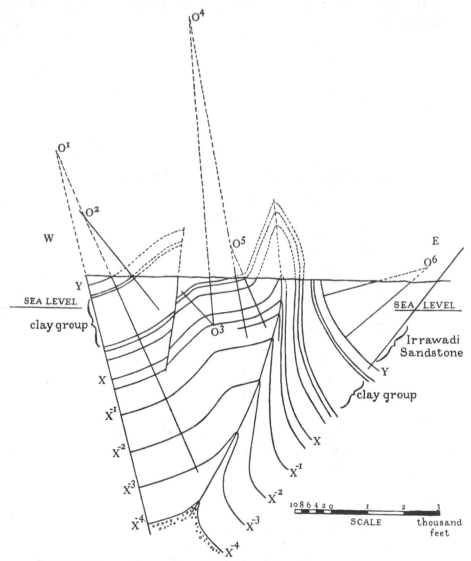

Fig. 62. Section across the Myaing anticline, Upper Burmah, showing a vertical and highly attenuated middle limb.

found with an unconformable junction, but, below this, the horizon lines can still be treated as lines of deformation. In this figure we add from a one-inch survey the dips whose normals meet at O^8, O^9, O^{10} and O^{11}. The axial plane becomes finally a hyperbola whose foci are at O^{11} and O^9. The auxiliary circle is very small, consequently the curve is almost a straight line, and its asymptote may be found within the limits of drawing by bisecting the straight line $O^{11}O^9$ at right angles. This section gives a good idea of the rapidity with which such a fold dies out at depth.

Myaing. The Myaing anticline is a good example of a fold with a vertical and highly attenuated middle limb, which passes further north into a reversed fault of great displacement. The outcropping horizons shown (Fig. 62) are sandstones, and it is the intercalated clays that become crushed out. The junction at horizon Y is disconformable, as the Irrawadi Sandstone facies lies above this. The conglomerates at X^{-4} crop out up

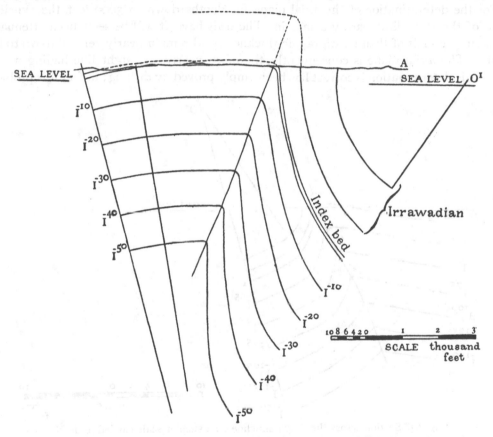

Fig. 63. Section across the southerly pitching end of the Singu anticline, Upper Burmah, showing an attenuated and nearly vertical middle limb. Within the limits of the drawing the axial plane is a straight line hading at 30 degrees.

the pitch. Attenuation is carried down from surface so as to decrease to nil, where the dips assume an angle of 50°. It is to be noted that the fold apparently begins to die out in the conglomerates below X^{-4}, which are very massive. The outcropping clay group is attenuated to a sixth of its normal thickness, and there must have been considerable movement along the bedding planes.

The gentle flank of the fold is competent, but, owing to the normal fault, cannot be drawn readily by tangential arcs.

Singu. Determination of curvature at depth of the important fold that constitutes the Singu oil field on the left bank of the Irrawadi is much handicapped by lack of evidence upon the steep flank. In the section (Fig. 63), taken from the southerly pitching end, there is no evidence that can be transferred to the section line beyond the point A, in which position the beds still remain vertical. The position of the synclinal bend is, therefore, conjectural and we can but place it in the most reasonable position, namely at or about the point O^1. In point of fact, from A eastwards outcrops are covered by an alluvial plain, and it is more than likely that the beds beneath this plain are gently dipping.

For the determination of the axial plane for depths down to 3000 feet, the synclinal bend of the steep limb hardly comes in. The beds have, it will be seen, been attenuated to about one-half of their stratigraphic thickness, and remain nearly vertical down to this depth. The axial plane is consequently, therefore, nearly a straight line hading at 30°. That this determination is correct has been amply proved by drilling; in fact it is probable

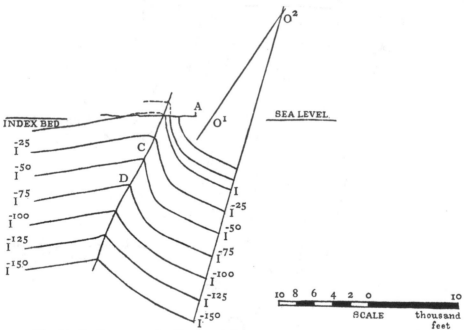

Fig. 64. Section across the Singu anticline on a smaller scale carried to depth.

that a more correct determination of the axial plane by a careful survey of the amount by which the middle limb has been attenuated, would have saved a considerable number of wasted holes over that part of the field which lies near the axis.

Assuming that the synclinal bend at O^1 is nearly correct, we may carry the fold on a smaller scale to depth (Fig. 64), and, by drawing in horizons, or lines of deformation at 2500-foot intervals, we may show that from about C to D there is an increase of westerly hade, as attenuation begins to die out, and below D, where there is no attenuation, the curve begins to bend back towards the vertical. Assuming that the gentle limb dips at about 10° for some distance beyond that shown, the final form of the axial plane is a parabola with O^2 as focus.

In the Yenangyat portion of the fold across the river (Fig. 58), the steep limb becomes very much more highly attenuated, and some authorities consider that there is here a reversed fault. This, however, would have the same effect on the axial plane as attenuation. The width of the steep limb here exposed is very small, and there is no evidence upon the left bank of the river, consequently our data are very limited. The greater degree of attenuation would indicate a general steepening of the axial plane at surface, and a more pronounced curve westwards at depth*.

From this cursory examination of some of the more important Burmese folds, it is clear how necessary it is to carry mapping far out into the adjacent syncline on the steep side. Unfortunately it is usually on this side that the evidence is missing. When this is so, we must do the best we can with what is exposed, bearing in mind that it is often the case that, where an alluvial plain obscures the outermost dips of the steep limb, it is close along the boundary of that plain that the synclinal bend will occur.

PROBLEMS CONCERNING THRUST ROCK SHEETS AS ILLUSTRATED IN SOUTH-WESTERN PERSIA

South-western Persia, in the foothill region between Dizful, Ram Hormuz and Bushire (Fig. 65), affords interesting evidence regarding the genesis and behaviour of thrust rock sheets, which, although occurring on a smaller scale and under conditions which are peculiar to the region, nevertheless throw considerable light on the mechanics of mountain structures of this type.

The region under consideration is flanked to the north-east by the great mountain ranges of the Zagros arc, which consist mainly of folded and thrust limestones of Cretaceous and Eocene ages, which are over-ridden still further to the north-east by a rock sheet consisting of older schists and limestones. On the south-west side there is the foredeep of Mesopotamia and the Persian Gulf. In this intermediate region occurs a series of rocks of such a heterogeneous character as to offer very differing degrees of resistance to the earth movements which affected them.

The small-scale geological map (Fig. 66), upon which no contours are shown, gives no picture of the ruggedness and inaccessibility of the region. The Kuh-Kinuh, rising to over 10,000 feet in the north-east, is separated from the Kuh-i-Salin by a region of such topographic chaos that it is rarely penetrated by the natives, and certainly no European has ever set his foot therein. Only a shade more accessible is the foot of Kuh-Kinuh

---

* Cf. Pascoe, *Oilfields of Burmah*, Plate 31. It is to be noted that in his sections across both the Singu and Yenangyat portions of the field, Pascoe does not allow for attenuation. His representation of the fold in these sections is consequently erroneous.

itself by a 5000-foot pass around the flank of Kuh-i-Hola, and though the oilfield region at the spot marked "Masjid-i-Sulaiman" on the map is now thickly populated, the very large sums of money that had perforce to be expended by the Anglo-Persian Oil Company on its thirty-mile road from the plains, testifies to the fact that, though the elevations are lower, topographic obstacles are still formidable.

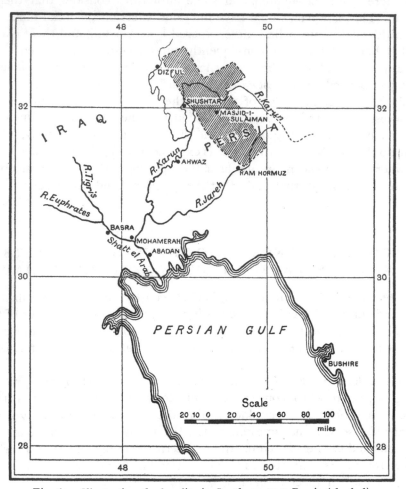

Fig. 65.   Illustrating the locality in South-western Persia (shaded)
here under discussion.

This rugged relief is due entirely to the fact that the region is geographically a very young one; the streams have nowhere approached grade, and each tectonic feature is accompanied by its corresponding topographic form. We see before us, in point of fact, a region which is being built up at the present day, and it affords unparalleled opportunity for studying tectonic forms in the process of their creation, and their contemporaneous degradation at the hands of the denuding agents.

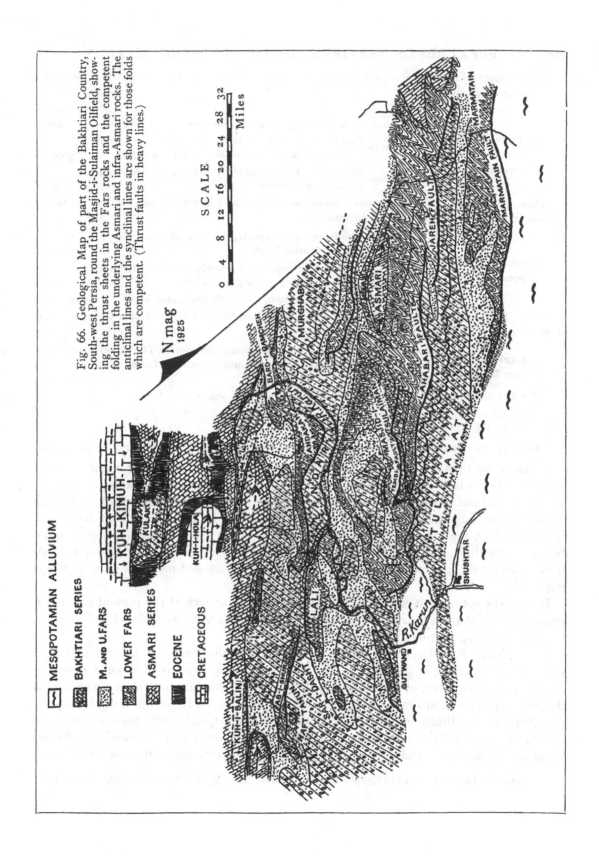

Fig. 66. Geological Map of part of the Bakhtiari Country, South-west Persia, round the Masjid-i-Sulaiman Oilfield, showing the thrust sheets in the Fars rocks and the competent folding in the underlying Asmari and infra-Asmari rocks. The anticlinal lines and the synclinal lines are shown for those folds which are competent. (Thrust faults in heavy lines.)

MESOPOTAMIAN ALLUVIUM

BAKHTIARI SERIES

M. AND U. FARS

LOWER FARS

ASMARI SERIES

EOCENE

CRETACEOUS

N mag
1925

SCALE

0 4 8 12 16 20 24 28 32
Miles

KUH-KINUH

KULANG

KUH-I-HOLA

KUH-I-SALIN

AB-I-SHUR

SAR-I-DASHT

TAFT

NILAB

LALI

ANDAKA

MASJID-I-SULAIMAN

NARUN

MURGHAB

BAID-I-QAMISHEH

ASMARI

TULLI KAYAT

LAHABARI FAULT

JAREH FAULT

MARMATAIN FAULT

MARMATAIN

GUTWAND

R. Karun

SHUSHTAR

A synopsis of the stratigraphy in the form of a table is first necessary. The geological succession is as follows:

			Feet
BAKHTIARI SERIES (Pliocene)	Upper Group	Massive conglomerates, sometimes	5000
	Middle Group	Intermediate type, river gravels, silts, clays, conglomerates, sometimes absent, and sometimes	5000
	Lower Group	River silts, clays, sandstones, with intercalated congl. bands, sometimes	5000

Note. The Bakhtiari Series assumes a deeper water facies as it is traced south-east. In the latitude of Bushire the Lower Group is largely marine.

FARS SERIES (Upper Miocene)	Upper Group	Sandstones and clays, largely of brackish water origin, a few marine bands	2700
	Middle Group	A passage group of clays, sandstones, limestones of detrital origin, a few gypsum bands	600
	Lower Group	Clays and gypsum, a few thin limestones up to	5000

Note. The Fars Series varies in the same sense as the Bakhtiari Series, the Upper Group assuming a limestone facies in the latitude of Bushire.

ASMARI SERIES (Lower Miocene)	Asmari limestone, a massive foraminiferal limestone. The upper part of this series sometimes assumes a passage facies with gypsum bands	1000
EOCENE SERIES	Foraminiferal (nummulitic) limestone and shales.	
CRETACEOUS SERIES	Massive limestone with hippurites.	

The general outline of the geological history of the area may be summed up as follows:

At the close of Asmari times (Lower Miocene) there was a shallowing of the sea, arms or lagoons of which were eventually isolated by contemporaneous warping of the sea floor. A land mass of limestone gradually rose to the north and north-east, where the main Zagros mountains now stand, and material from this, together with local contributions from the warped up promontories, gave rise to the fine detritus, which is interbedded with the Lower Fars gypsum. The isolation of the highly saline Lower Fars sea must have been completed early in the period, and gypsum and rock salt were copiously deposited.

It is worth noting that rock salt rarely appears at the surface at the present day, but is found in considerable masses in the wells drilled by the Anglo-Persian Oil Company. An adequate explanation for this is still lacking *.

The Middle Fars Group represents a "passage" period between the saline Lower Fars, and the marine or fresh water Upper Fars. This group varies to some extent in lithology over large areas, the marine type generally predominating at the expense of the gypsum to the south-east. Taking individual areas in the region as a whole, deposition, both in the Middle and Upper Fars, was very regular, and thicknesses are uniform, but in the Upper Fars particularly there is a change from marine conditions in the south-east to fluviatile conditions in the north-west, and this regional variation is main-

---

* Similar conditions are found in Egypt: see W. F. Hume, *The Oilfields Region of Egypt*, Cairo, 1916.

tained through Bakhtiari times. Thus fluviatile or coastal sands of the Upper Fars of one particular horizon in the north-west may be represented by marine limestones in the south-east, while gravels or pebble beds of the Lower Bakhtiaris in the north-west may be represented by river silts and clays at the same horizon in the south-east. There was in point of fact a coast line which advanced from north-west to south-east during the Middle and Upper Fars and Lower Bakhtiari periods.

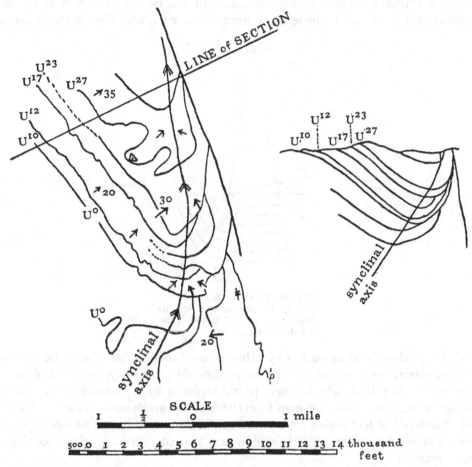

Fig. 67. The southerly pitching end of the Lahabari syncline showing the effect on the map of the "counter-hade" of the axial plane, brought about by earth movements contemporaneous with deposition. Horizon indices refer to the number of hundreds of feet above the base of the Upper Fars Group. The effect of "counter-hade" may also be noted in the section.

Although in the main the thicknesses of the Upper Fars Group are remarkably constant, there was during this period in some places, and along the lines of what were to become the main axes of folding, an initiation of those movements, which were to continue with only one pause down to the present day. During the laying down of the whole thickness of the Bakhtiari Series, which in some of the synclines reaches 15,000 feet, these movements operated contemporaneously with deposition, giving rise to an infinite number of unconformities and overlaps, to which we have collectively given the name "continuous unconformity." The result of this simultaneity of movement and deposition was a thinning out of the series along the flanks of the various established synclines (Figs. 69, 70). This is often so marked as to give rise to what we have called a "counter-hade" of the axial plane of the syncline. In a normal or competent syncline, the axial plane hades towards the steep limb (Fig. 68); but in a syncline whose steep limb is uplifted contemporaneously with the deposition of the beds of which it is formed, the axial plane will be found to hade away from the steep limb. Fig. 67 shows the actual

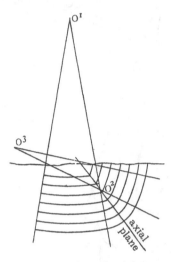

Fig. 68. To show that in a competent asymmetric syncline the hade of the axial plane is towards the steep limb.

map of the pitching end of such a syncline, where "counter-hade" may be detected in the surface plane; for it will be noted that, as the fold pitches up, the axial plane moves away from the steep limb, which is here partly replaced by a reversed fault. The figure also shows a section across the map from the data shown thereon. Horizon indices are given in hundreds of feet above U^0, and indicate normal thicknesses as developed to the north-west along the gentle limb. It will be seen that counter-hade of the axial plane is very well marked.

Deposition and contemporaneous folding continued throughout Bakhtiari times, and depression of the synclines became more rapid, though deposition more than kept pace with the movement. The synclines eventually became filled up with gradually coarsening material, and at last the conglomerate of the Upper Group began to encroach on the anticlinal areas as well. As soon as this invasion had become general, the old consequent drainage lines along the synclines altered in direction, and material was carried direct from the limestone mountain ranges to the north-east and deposited in a number of vast gravel cones by the rivers, which now debouched upon a single great gravel plain. The close of the period, therefore, was characterised by mountains to the north-east, and a gravel plain to the south-west, which effectually hid the north-west south-east trending folds in the Fars and Lower Bakhtiaris beneath them (Fig. 71).

DIAGRAMMATIC SECTIONS ILLUSTRATING THE HISTORY OF SOUTH-WESTERN PERSIA FROM ASMARI (LOWER MIOCENE) TIMES TO THE PRESENT DAY

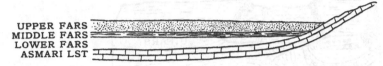

Fig. 69. The deposition of the Lower, Middle and Upper Fars against the Iranian continent under mainly quiescent conditions.

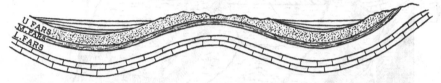

Fig. 70. Part of the above section, showing the deposition of the river deposits of the Lower Bakhtiari with contemporaneous flexuring during deposition.

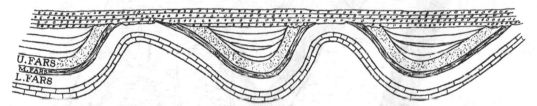

Fig. 71. The whole area covered with the gravels and boulder beds of the Upper Bakhtiari, with temporary quiescence.

"GAMMA" STRUCTURE   "OMEGA" STRUCTURE

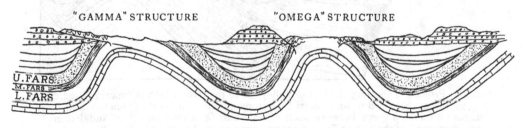

Fig. 72. Degradation of the Upper Bakhtiari Plain, and re-exposure of the Lower Fars gypsum on the crests of the anticlines; illustrating the thrusting which now took place in these beds.

There was then quiescence for a time, though probably no slackening of the forces which produced the former folding movements, for the thick covering of rigid conglomerates would have been an adequate protection against the continuation of such movements. It is unnecessary also to invoke regional uplift to account for the dissection of the gravel plain, which now followed. This would certainly have come about in the normal sequence of events as the rivers became less loaded. However that may be, dissection followed in a very haphazard fashion, along any drainage lines that happened to exist on the plain at the time. All the present-day main drainage lines from the mountains to the plains are superimposed, though there has been a working back along the strike,

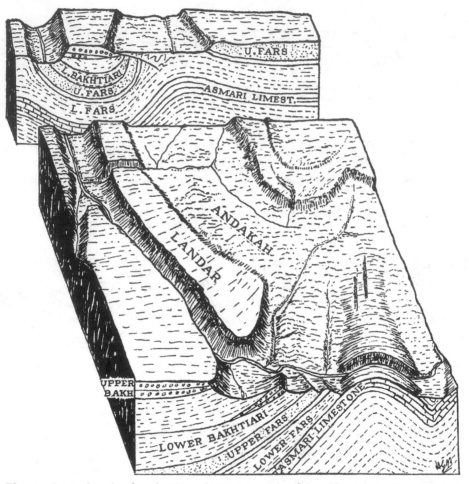

Fig. 73. Projection showing the Karun Valley where the river, which is superimposed, cuts through the pitching end of an Asmari Limestone anticline at Bard-i-Qamcheh and then plunges into a deep gorge between walls of Bakhtiari Conglomerate. The Andakah rock sheet, composed of the Lower Fars, is shown invading the Landar syncline. The drainage of the Andakah plain escapes into the Karun River by a valley which hangs 1200 feet.

where the softer rocks are now uncovered. The River Karun itself could hardly have chosen a path more difficult to erode. It debouches into our region at Bard-i-Qamcheh, makes straight for a great wall of conglomerate (Figs. 66, 73), where it is most thickly represented, and then plunges through a gorge in a series of great cascades completely out of grade with the main part of the stream. Turning north-west it moves more quietly in a number of incised meanders, then, carefully avoiding a course, which might have led it to the plains direct through the easily eroded gypsum, keeps to its entrenched conglomerate valley, and debouches on to the plains by a deeply incised canyon at Gutwand. Below this place, as an antecedent stream it cuts through two ranges that lie in its path, the Tul Kayat range at Shushtar, and the Ahwaz range at Ahwaz some 70 miles further down. In fact it follows the old course that it originally held over the old plain of Bakhtiari conglomerate, at that point of time immediately prior to its period of incisement.

Figs. 69, 70 and 71 illustrate the general history up to the close of Upper Bakhtiari times, which probably range high up into the Pliocene, and it is during the next phase, after the incisement and degradation of the Upper Bakhtiari plain and the exposure of the Lower Fars gypsum along the old anticlinal lines, that we begin to get a higher degree of complication in the structure.

From the stratigraphic table we may note that we have now a thickness of some 5000 feet of gypsum bedded between two thick rigid groups, the Asmari Limestone and infra-Asmari

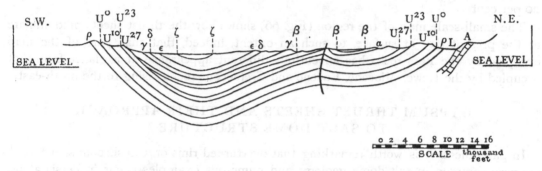

Fig. 74. Section across the Murghab syncline, Persia, drawn true to scale, showing the rate of variation in a highly variable series of river deposits, with time horizons a to ζ. The section also shows the rate of variation in a highly variable series of river deposits, with time horizons a to ζ. The section also shows the axial plane of a small fold, which, hading south-west at surface, bends through the vertical and finally hades north-east in consonance with the general dip of the north-eastern limb of the syncline. The comparative rates of uplift of either limb may be judged from the comparative rates of overlap and thinning out. Movement on the south-western limb, it may be seen, was the more rapid.
A, Asmari Limestone; L, Lower Fars; ρ, Middle Fars; U^0—U^{27}, Upper Fars; a—ζ, the highly variable Bakhtiari Series.

beds below, and the Upper Fars and Bakhtiari Series above. Now gypsum under quite moderate stresses acts as a plastic body. Under vertical stresses, brought into play by overlying relict mountains of Bakhtiari conglomerate, it tends to bulge out along lines or at points of weakness in the superincumbent strata into domes, like those of Transylvania, while under horizontal stresses it readily develops thrusting.

As denudation began to expose the gypsum where it was nearest the surface, that is along the crests of the buried anticlines, overfolding and thrusting of these pliant rocks immediately began to operate, generally from north-east to south-west, though there is frequently back thrusting in the opposite direction as well. Thus the Lower Fars Group began to be thrust forward bodily over the steep limb of each foresyncline in turn. In the process the gypsum itself became highly folded; at the front of each advancing sheet it often became rolled and inverted again and again, while further behind it took up the form of highly compacted isoclinal folds with nearly vertical axes.

Fig. 72 illustrates the general result of this process, and Fig. 75, a section across part of the Maidan-i-Naftun oil field drawn accurately and to scale, shows a thrust sheet which has advanced over and covered the whole of its foresyncline.

The mechanics of these thrust movements are not always easy to decipher. Although there is generally sufficient evidence to calculate the area in cross section of the thrust sheet, or of its total volume, it is not so simple to determine exactly whence this thrust mass has come, owing to the fact that the attenuation, which the middle limb has suffered by the squeezing outward and upward of the Lower Fars, is always obscured by the thrust sheet. In the section shown (Fig. 75) the total stratigraphic thickness of the Lower Fars is about 5000 feet, for the position of the sub-thrust outcrops of the Middle Fars has been found by mapping the pitching ends of the foresyncline, and the position of the limb of the Asmari Limestone is given from the evidence of the drilled wells. It is fairly clear from this that the bulk of the thrust material has been sheared off the north-eastern side of the fold. Attenuation of the gypsum of the middle limb, it will be noted, is by about 20 per cent.

The small-scale map of the region (Fig. 66) shows how the thrust sheets predominate at the present day at surface, to such an extent, indeed, that that part of the map occupied by the Fars rocks has an entirely different appearance at a glance from that occupied by the Lower Miocene, Eocene and Cretaceous rocks, rising to the north-east.

GYPSUM THRUST SHEETS AND THEIR APPROACH TO SALT DOME STRUCTURE

In parenthesis it is worth remarking that overturned rims or "mushroom structures' are very common in salt dome geology, and numerous examples occur in Persia along the Bandar Abbas-Lingeh seaboard. It has been shown in this region that the salt is of great age (Pre-Cambrian) and of unknown, but probably great, thickness, and it is clear, from the general disposition of the salt domes, that the salt, which is demonstrably plastic under pressure, becomes extruded at specially weak points, generally in circular domes, from which the overlying rigid sediments have been removed, and along lines of weakness, generally the axes of anticlines. It can be shown with fair certainty as well, that, where only a small area of salt has been exposed, possibly by only a comparative pin-prick through the superimposed strata, that extrusion becomes rapid, and that the salt behaves like a semi-liquid or viscous mass, and, lapping over the surrounding rocks, overturns these beds in the process. Extrusion may thus only be, in its final and more obvious phase, the result of pressure upon the underlying salt surrounding the dome.

Gypsum behaves very similarly under similar conditions though in a slightly less degree, and its plasticity can be demonstrated in the field. All the great thrust sheets in Persia are composed almost entirely of what may be called by the analogy of its micro-structure *gypsum quasi-schist* (Fig. 74). This rock consists in the main of gypsum with inclusions of unaltered clay shale and limestone locally derived. The gypsum appears to have flowed round the inclusions. The rock is more than a fault breccia, as it may occur many hundreds of feet thick, and the whole of a thrust sheet may in fact be composed of

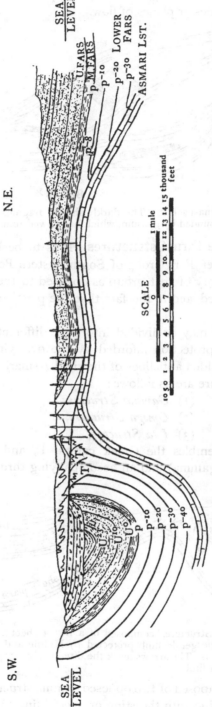

Fig. 75. Section across the Masjid-i-Sulaiman oil field showing a thrust sheet of Lower Fars rocks overriding nearly the whole of the foresyncline. The figure also illustrates the quantity of material sheared off the gentle limb, and the two-mile displacement of the upper fault block across the Asmari Limestone. Note that there is nowhere contact between the rigid Middle Fars and the Asmari Limestone, and that there is a five hundred foot thickness of the gypsum and clays of the Lower Fars Group, interposed between the two blocks, which acts as a kind of lubricating medium between them.

it. In a group such as the Lower Fars, it is not always possible to determine whether we are dealing with bedding planes or planes of flow.

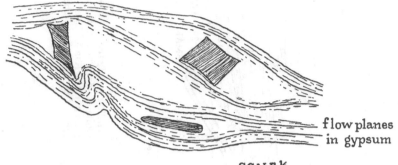

flow planes
in gypsum

SCALE ¼

Fig. 76. "Gypsum Quasi-schist." The darkly shaded fragments are shale or limestone surrounded by gypsum, which has flowed round them.

The classification of the thrust structures. It is to be borne in mind that the thrust structures in the Lower Fars Group of South-western Persia owe their existence entirely to the greater plasticity of that group as opposed to the rigidity of those rocks between which it is sandwiched, and to the fact that the plastic rocks, being exposed at surface, were free to move.

The structures which occur may be divided into three different forms dependent upon various factors, such as the protection afforded by the overlying rigid strata, and the proximity and form of the hidden anticlines of the rigid Asmari Limestone.

The three classes of structure are as follows:

(1) *Gamma Structure,*
(2) *Omega Structure,*
(3) *Iota Structure.*

(1) *Gamma Structure* resembles the Greek capital Γ, and consists of a reversed fault (the vertical line to the gamma), and an accompanying thrust or overflow sheet (the horizontal line) (Figs. 77, 78).

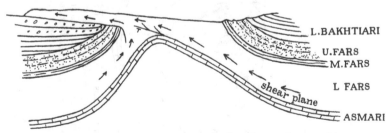

L.BAKHTIARI

U.FARS
M.FARS

shear plane   L FARS

ASMARI

Fig. 77. "Gamma Structure," composed of a thrust sheet and its roots. The figure shows the gentle limb protected by Middle and Upper Fars and Lower Bakhtiaris. The arrows mark the direction of movement of the Lower Fars up both limbs.

(2) *Omega Structure* is composed of two opposed gamma structures, ꓶΓ, and resembles the Greek capital Ω, an anticline with thrusting or overfolding to each limb (Fig. 77).

(3) *Iota Structure* is simple reversed faulting with no accompanying thrust sheet. It is a Γ without the horizontal stroke.

The Lower Fars thrust sheets, in either gamma or omega structure may be made up of a great number of small structures of a similar nature, which can often only be represented on a small-scale map diagrammatically by a number of isoclines. Each individual fold may, however, be mapped on a large scale, and Fig. 78 shows how a thrust sheet may be thus constructed.

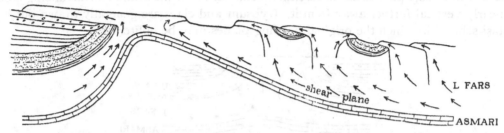

Fig. 78. "Gamma Structure" where the gentle limb is unprotected, showing minor gammas and omegas there developed in consequence. The arrows mark the direction of movement in the Lower Fars. In both this figure and in Fig. 77, a near approach of the rigid Asmari Limestone is postulated.

The intensity and complexity of the minor folding, associated with the great thrust sheets, are dependent upon the measure of protection afforded by the more rigid strata overlying the Lower Fars. If such protection is absent, the minor folding may be of any degree of intensity, but a protecting mass of a few hundred feet of sandstone and clays of the Upper Fars may be sufficient to prevent any such irregularities, and movement is taken up alternatively by the shearing out of great thicknesses of the Lower Fars against the Asmari Limestone (Fig. 77).

Omega or mushroom dome structures are very common throughout the region, though this structure may take up a number of variations. There is "linear" omega structure, where both limbs and even the pitching ends of a long anticline may be very steep, over-folded or thrust, and there is true domal omega structure where the form in plan of the upthrust gypsum may be circular with a thrust fault or an overfold completely round the circumference.

It may be stated in parenthesis that, although the terms *iota, gamma* and *omega* structure are intended to be applied only to this special region, these types of structure are often found in other parts of the world, where, for instance, poorly consolidated Tertiary clays, or other non-rigid rocks, are sandwiched between two groups of comparatively greater resisting material.

It will be seen from the figures shown that the type of structure taken up depends upon the proximity of the rigid Asmari Limestone beneath surface, and the measure of protection afforded by the superincumbent rigid Upper Fars and Bakhtiari Series. In the figures (Figs. 77, 78, 79) arrows indicate roughly the movement of the gypsum that has taken place. Thus, gamma structure generally indicates close proximity of the Asmari Series beneath, with either considerable protection afforded by the overlying beds (Fig. 77), when there need be little complication by minor structures of the thrust sheet, or where there is no protection, when a variety of gammas and omegas on a small scale form up behind each other on the gentle limb (Fig. 78). Omega structure may occur quite independently of any folding in the Asmari Limestone beneath, and may occur in any spot, where there is weakness in the protective covering of the later rocks, a small exposure of the Lower Fars often leading to the welling up of that group into an exaggerated omega (Fig. 79). The actual overflow thrust sheets themselves consist generally of closely packed isoclines, either recumbent near the front of an advancing block, or nearly vertical further away from it. Gypsum and clays are generally drawn out into "quasi-schist," in which the original bedding planes may be obliterated.

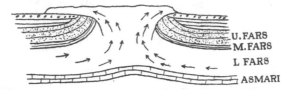

Fig. 79. "Omega Structure" composed of two opposing Gamma structures. Omega structure on a large scale does not imply a near approach of the rigid Asmari Limestone to surface.

There is ample evidence to show that these thrust movements have continued down to the present day, and are still continuing. Each advancing Lower Fars sheet is marked by an escarpment, which may be denuded to a greater or lesser degree, dependent upon the rate of advance of the thrust sheet wall. This becomes cut back by consequent streams, giving rise to triangular facets, each facet being part of the original face of the advancing wall (Fig. 80)*. Where the thrust movements are at the present day proceeding rapidly, there is always a steep face into which the consequents have incised themselves but little, and even where advance has not been particularly rapid, there is always a well-marked hanging valley system, by which the drainage from the top of the fault block is precipitated (Figs. 80, 83).

As may be well imagined, diversion of drainage by the advancing wall of a fault block is extremely common, and all the anomalies of a superimposed river system are thereby intensified. Lack of grade is in all the main streams commoner than any approach to a base line of erosion. One or two notable instances may be mentioned in particular. At Lali (Figs. 81, 82) the Ab-i-Shur has been forced to flow between the walls of two advancing and opposing thrust sheets, and has thereby found itself compelled to cut through a considerable thickness of massive Bakhtiari conglomerate. In doing so this active river has carved for itself a canyon 1200 feet deep, only 10 feet wide at the bottom, and under 200 feet wide at the top. Had the river succeeded in cutting its way out a little further to the south-west, or had there been capture from this direction, the

* Compare also similar topographic forms due to normal faulting, where, however, the escarpment wall is coincident with the fault face, and not as here nearly at right angles to it. W. M. Davis, *Geographical Essays*, 1909, p. 725.

Fig. 80. The advancing face of the Lahabari Thrust sheet, showing its hanging valleys. The rocks of the advancing wall are highly folded and sheared gypsum and clays of the Lower Fars. The plain in the foreground is a syncline.

valley would have traversed a considerably lower plateau of the more easily eroded gypsum.

Another fine example of a river, which, although maintaining its consequent course along the axis of a syncline, has been considerably constricted in its action by two

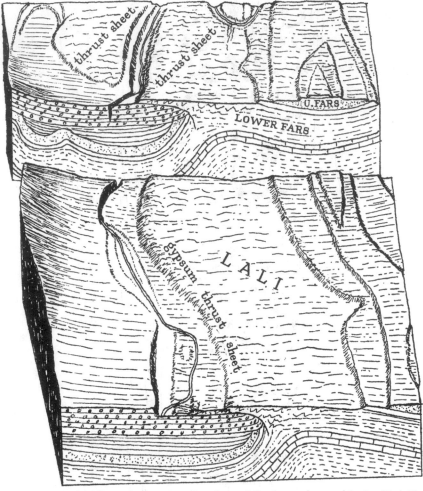

Fig. 81. The Lali Plain, South-western Persia, and its thrust sheet, with the Ab-i-Shur plunging into its canyon in the Bakhtiari Conglomerate. It has been compelled to maintain its course in the syncline composed of these hardly eroded rocks by the two advancing walls, the one of the Lali thrust sheet, and the other a thrust sheet encroaching over the syncline from the opposite direction. The length of the figure is about 5 miles.

advancing and opposed fault walls, is that of the Tembi River about 5 miles north-west of the Anglo-Persian Oil Company's pump station of that name. The advancing fault walls have approached one another to within a few hundred feet, the north-easterly fault sheet having encroached practically over the whole of the Tembi syncline (Fig. 83). The passage between the fault walls is only maintained by the activity of the river, and if this were at any time to be diminished, there would be a pounding back of its waters into a lake. This actually happens in the case of the Ab-i-Shur in flood time, where the width of the canyon is insufficient to take all the water that is brought down with the melting of the snows. The depth of this temporary and seasonal lake may be over 100 feet, and it has its own system of shore lines (Fig. 81).

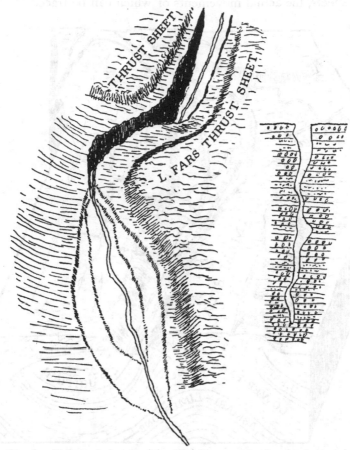

Fig. 82. Enlarged sketch of the Ab-i-Shur crack, showing how the river has been compelled to follow a course by which it has to cut through a great thickness of Bakhtiari Conglomerates, by the advancing walls of two opposing thrust sheets. A section through the crack, which is 1200 feet deep, drawn true to scale is shown. Above the crack are shown the terraces and shore lines of the temporary lake, that occurs at flood time, when the flood waters become constricted in the crack.

In structures such as those we have outlined geometric constructions for section drawing should be applied as far as possible. Where there is much attenuation and variation this may be difficult, but large segments of the successive foresynclines are often competent, and are often susceptible to direct geometric methods, and it is in their correct mapping and delineation beneath surface that the calculation of thrust displacement depends. Much of the Tembi syncline (Fig. 75) has been constructed geometrically from the surface evidence, as exposed in those localities where the thrust sheet has not encroached to the extent that is shown in the figure. Correct mapping in Persia has been entirely dependent on checking thicknesses by geometric construction, as the work proceeds.

South-western Persia thus affords valuable evidence regarding the behaviour of advancing rock sheets, the actual movements of which can be traced at the present day.

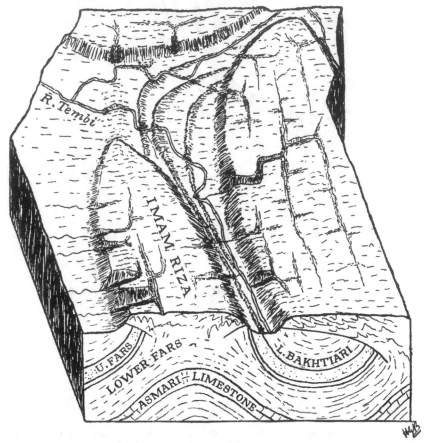

Fig. 83. Projection showing the north-westerly pitching ends of the Tembi syncline and the Masjid-i-Sulaiman anticline, the thrust sheet of the latter having advanced over and covered almost the whole of the former. The Imam Riza is an omega structure. Note the valleys which hang over the fault escarpments. The River Tembi barely succeeds in maintaining its course between the two opposing fault walls. The length of the figure is about 6 miles.

The region is, in point of fact, a working model of tectonic process, on rather a small scale perhaps, but subjected to relatively small stresses, and acting over a relatively short period of time. If we compare one of the Lower Fars rock sheets and its environment, with one of those far greater rock sheets of the Alps, we find that many essential features are present in both. There is the rigid block in front, in Persia always a syncline, and the roots of the fold behind. There is the minor folding of the thrust sheet itself, and occasional back folding in both (omega structure). But the rock sheets of the Alps are in point of fact a series of gamma structures*.

Persia teaches us that there is much to be learnt regarding the relative resistance of the various rocks to earth stresses, for the thrusting is there confined to the easily folded and pliant gypsum and clays of the Lower Fars.

RIFT VALLEY STRUCTURE AS ILLUSTRATED BY THE EASTERN MARGIN OF THE GULF OF SINAI

The structures associated with those parts of the earth's crust, which have been directly subjected to tensional stresses, are in marked contrast to those which have been under, and given way to compressional strains. In the latter, folding and displacement by thrust and tear faults are the main features, and normal faulting may be absent or only subsidiary; in the former, folding is subsidiary, and there is rarely if ever true thrusting, while normal faulting is the main feature.

The Great African Rift Valley is one of the most striking features of the earth's surface, and occupies in length about one-sixth of its circumference. Only widely separated districts of the region have as yet been studied in any detail. It is natural, therefore, that the structural types met with are found to vary considerably. The Gulf of Sinai, a branch of the Great Rift Valley, and associated with the subsidences of the Red Sea and the Gulf of Akabah, on account of its easily followed stratigraphy, affords us an excellent lesson in strike faulting and associated flexuring.

The peninsula of Sinai, though strongly folded in the north into a number of domes with an east and west orientation, rises as a gently dipping fault block from the latitude of Suez and Akabah, Miocene, Eocene, Cretaceous and Archaean rocks being exposed in succession to the south. There is also a topographic rise from this latitude, where the plateau stands at about 1500 feet, to the deeply dissected region of Mount Serbal, which rises to 6000 feet.

Although the study of the area, which is typically a rift fault region, is contrasted with that of compressional flexuring in that folding occupies only a subsidiary place, the rocks as a whole being left horizontal or gently dipping, none the less the main faults, which build up the Sinaitic fault block and its adjacent rift valley, are rarely simple, and there is subsidiary warping and folding, which brings about the entrance and dying away of each individual fault in échelon.

* Cf. Alpine literature generally, e.g. L. W. Collet, "Geology of the Swiss Alps," *Proc. Geol. Assoc.* XXXVII, p. 375, Fig. 39. Also by the same author, *The Structure of the Swiss Alps*, 1927. This book embodies all the latest and the most authoritative work on the Alpine Thrust Sheets.

The folding of the Lower Fars Group of South-western Persia has been worked out independently and from different data, hence the terminology there adopted is retained. In any case the English translation of foreign terms in tectonic geology is usually preferable, as "thrust sheet" for "nappe"; "upthrown block" for "horst"; "downthrown block" for "graben," and so on.

The stratigraphic succession, so far as it concerns the Upper Cretaceous and Eocene of the Sinaitic margin of the Gulf of Suez is extraordinarily regular, and thus renders the calculation of fault throws particularly easy. There is considerable variation in the whole succession in a north-easterly direction towards Palestine, both in thickness and type, but with this we need not concern ourselves.

A summary of the succession and its thickness on the coast is as follows:

			Metres
MIOCENE SERIES		Clays, shelly grits, gypsum (Thickness variable)	
		Unconformity with basalt lava flows from dyke fissures.	
EOCENE SERIES	Upper Group	Marls, limestone and gypseous marls. More than	310
	Lower Group	Massive chalky limestone weathering brown ...	250
UPPER CRETACEOUS	Maestrichtian	Grey shales. 70 m.	160
		Chalky flagstones with Ostrea vesicularis. 90 m.	
	Santonian	Clays and variegated marls with Echinobrissus waltheri	150
	Turonian	Massive chalky limestone with Leoniceras segne	100
	Cenomanian	Clays, shelly limestone marls with Ostrea mermeti and Exogyra flabellata	100
LOWER and INFRA-CRETACEOUS	Nubian Sandstone	In general massive quartzose sandstone (Thickness variable)	
		Unconformity.	
ARCHAEAN COMPLEX		Gneisses, schists, etc.	

It will be seen at once from the general map (Fig. 84) that there is no single fault line bordering the Gulf of Suez, but a number of major structures arranged en échelon, and in two roughly parallel bands, which may be termed respectively the inner and outer Sinaitic system. The throw of the various faults which compose the outer Sinaitic system is somewhat masked by deposits along the sea coast, but on the other hand the great regularity in thickness of the Lower Eocene (250 m.) and the Upper Cretaceous (510 m.) from north-west to south-east renders the task of calculating fault throws somewhat easier.

The faults may be divided into two main types depending upon the flexuring by which they come in and die out.

(1) **The crescentic fault** type (Fig. 85) depends for its form on the fact that the upthrust block dips away from the fault face. Thus any change of strike in the direction of the dip of the upthrow block brings about a diminution of throw. The fault dies away in either direction by a change of strike in this sense. The figure shows an upthrow block and its fault scarp. There has been erosion of the upthrow block, which has exposed some of the lower beds.

Fig. 86 is a map of a fault escarpment, with a crescentic fault (the Feran Fault), whose actual outcrop is here obscured by alluvium. The beds on the downthrow side are also obscured, but are shown further south to be of Miocene age. It will be noted that the upper fault block dips away in a general east-north-easterly direction, and that the fault bends round to the north-east just to the north of the "Govt. Well Location." There is conse-

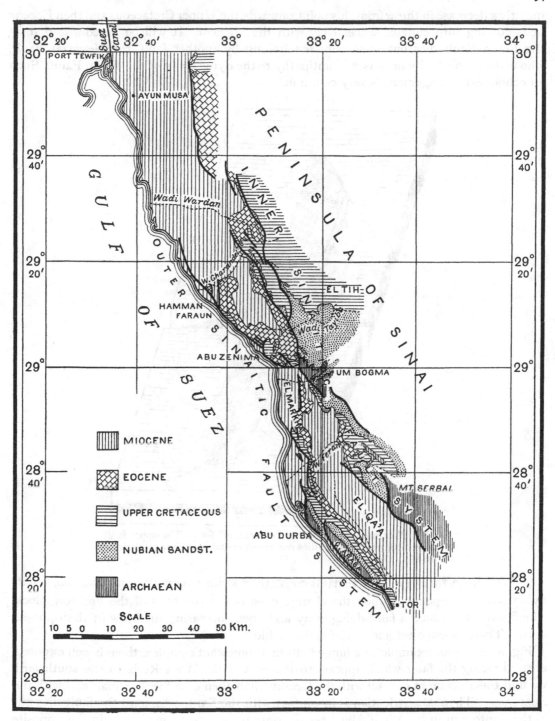

Fig. 84.

quently a decrease in throw from this point onwards, the Upper Cretaceous and then Eocene beds coming into contact successively with the Miocene. It will be noted also that, as the Feran Fault dies out, movement is taken up by another fault, the Nezzazat Fault, whose throw gradually increases in antipathy to the dying away of the Feran Fault. Such an echeloned arrangement is very common.

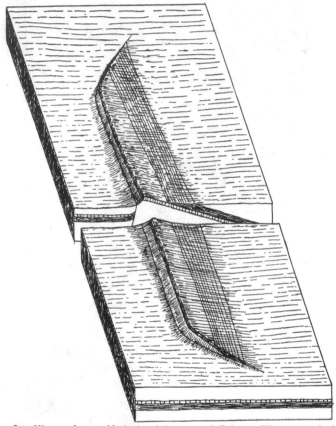

Fig. 85. Illustrating a rift fault of "crescentic" form. The upper fault block is tilted and the fault dies out at either end by a change of strike in the fault itself.

(2) **A hinge fault** (Fig. 87) remains constant in strike over most of its distance, and its dying away is dependent upon the change of strike of the beds of the upthrow block. The figure shows such a fault dying away and then increasing in throw in the opposite sense. There is some erosion of the upthrow block.

Fig. 88 gives an example of a hinge fault in a somewhat complex, though well exposed, area. Tracing the fault which appears to the east of the Wadi Resiz to the southward, we find the Eocene in contact with the Cenomanian, then with the Turonian and with the Santonian. Then Maestrichtian is in contact with the Santonian, until the fault dies out in the Santonian in the Wadi Abu Lesseifa, only to re-enter immediately in the opposite

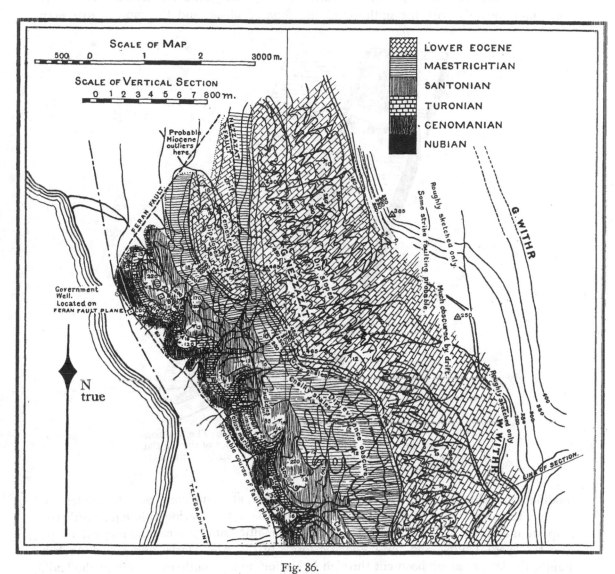

Fig. 86.

sense with the upthrow side to the west. The fault increases gradually in throw but dies out again in Turonian rocks west of point 525.

Passing to a consideration of the Gulf subsidence in general, we may gather from Fig. 89 the general arrangement of the lines of movement. Erosion, it must be remembered, acted contemporaneously with the upthrow of the Sinaitic block, and the fault scarps were

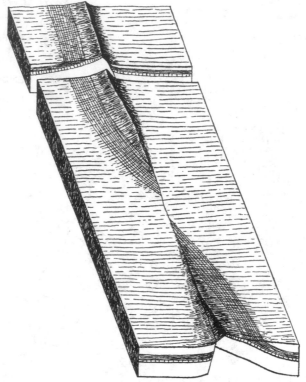

Fig. 87. Illustrating a rift fault of "hinge" form. The upthrow and downthrow sides interchange. The dying out of the fault at the hinge is brought about by a bending of the strike of the beds of the respective upthrow blocks.

modified accordingly. Uplift has proceeded, however, at a sufficient pace to render the tectonic features easily recognisable at the present day. Comparing this figure with the map, we may trace both the inner and the outer Sinaitic fault systems, the hinge fault, *H*, near the Wadi Tayiba, and the crescentic forms of the Abu Durba and Gebel Arabah Faults, the latter having been cut through in section by the southern margin of the fault.

The faults shown in the figures are necessarily idealised. There are often minor complications, minor irregularities in throw and replacement of the main fault by a number of step faults.

Rift faulting of necessity affects the crust of the earth more deeply than regional folding. In fact, it is uncertain to what depths the great rift faults may be not carried. Such move-

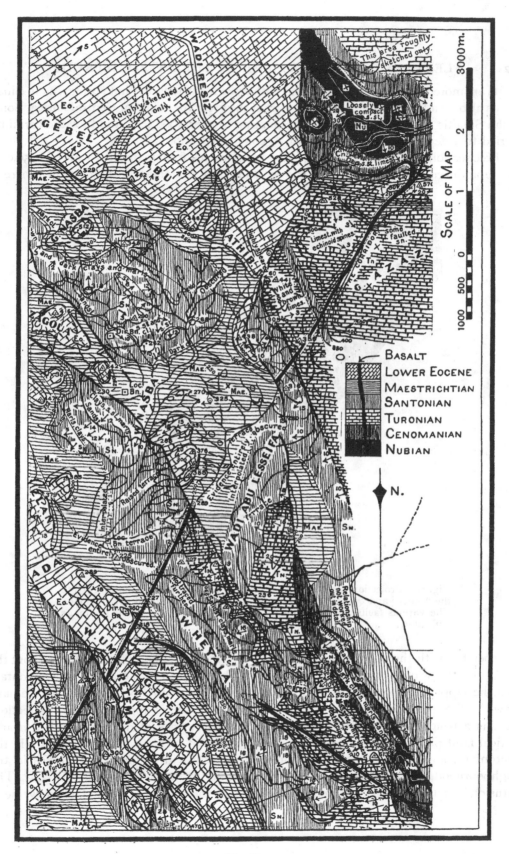

Fig. 88.

ments are more often than not associated with dyke intrusions and hot springs, as in Sinai, and as also on a larger scale in the main part of the great rift valley of Central Africa. Some of the Sinaitic faults involve a throw amounting to over 5000 feet, and they are certainly carried to a depth beyond the limits of a geological section.

Three sections are shown (Figs. 90, 91, 92) constructed from maps on a scale of 1 : 50,000, and reproduced here on that scale. Fig. 90 shows the Hammam Faraun Fault, where it

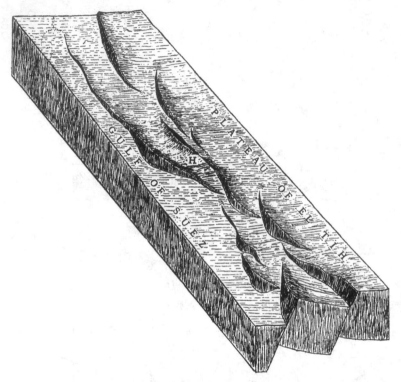

Fig. 89. Projection showing the rift faults of the Gulf of Suez margin. Note the "inner" and "outer" systems, and the general echeloned arrangement. The various faults can be identified from the map, Fig. 84. Note the hinge fault at *H*.

is associated with the hot springs (about 120° F.) of that name. There is doubt about the maximum throw of this fault, as it is uncertain what horizon in the Miocene is in contact with the Upper Cretaceous, for the downthrow side is obscured by the Gulf, but we have shown in the section a minimum throw. As the fault line is approached in the section, there is a bending over of the beds, and such folding, associated as it is with a nearly vertical fault plane, must be similar and not competent. The section shows a sill in the Santonians, an intrusion, which is connected directly with a number of dykes in the neighbourhood, which in turn feed a lava flow that occurs at the base of the Miocene. The intrusion is thus dated exactly, and probably marks the first stage in the fault movement.

Fig. 91 is again taken from a map on a scale of 1 : 50,000. It shows two faults that are in échelon, the Markha Fault taking up part of the movement of the Darat Fault, which is dying out to the south-east. There is an intervening folded syncline, which is probably competent. The throw of the Markha Fault can be measured exactly, but a minimum is shown for the Darat Fault, as the horizon exposed of the Miocene rocks on the downthrow side is uncertain.

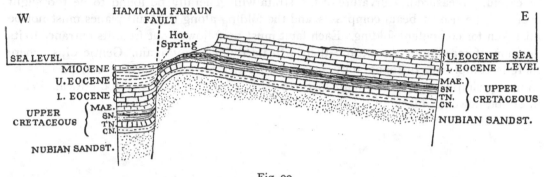

Fig. 90.

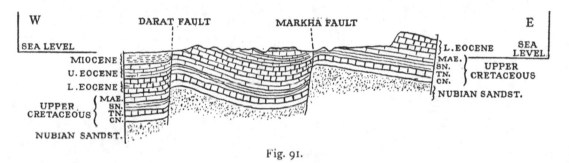

Fig. 91.

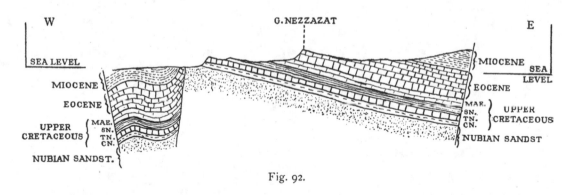

Fig. 92.

Fig. 92 shows the Feran Fault in a position about 6 km. to the south of the "Govt. Well Location" (Fig. 84). The throw of the fault is exact, and the folded downthrow block is exposed at surface. The upthrow block of Gebel Nezzazat is a "monoplane." The fault is of crescentic type, and dies out by changing its strike to the east, thus bringing higher and higher beds in contact with the Miocene on the downthrow side.

Though the problems met with in block faulted areas such as this are very different from those presented in folded areas, their solution by geological mapping follows very much the same principles. As before, a contour map is required, wherein thicknesses can be carefully measured. Curvature of the strata will generally be found to be too slight even for the use of beam compasses, and the folding along the fault planes must not be mistaken for competent folding. Each fault must be followed out from its entrance to its position of maximum throw, and thence to its dying away again. Gentle dips, where there is complexity of topography, mean complexity of outcrop, which may perhaps be shown as simplified upon the map, in those parts of it which are not too critical. When in doubt, however, fill in all the detail.

INDEX

Printed in the United States
By Bookmasters

Printed in the United States
By Bookmasters